Springer Theses

Recognizing Outstanding Ph.D. Research

Aims and Scope

The series "Springer Theses" brings together a selection of the very best Ph.D. theses from around the world and across the physical sciences. Nominated and endorsed by two recognized specialists, each published volume has been selected for its scientific excellence and the high impact of its contents for the pertinent field of research. For greater accessibility to non-specialists, the published versions include an extended introduction, as well as a foreword by the student's supervisor explaining the special relevance of the work for the field. As a whole, the series will provide a valuable resource both for newcomers to the research fields described, and for other scientists seeking detailed background information on special questions. Finally, it provides an accredited documentation of the valuable contributions made by today's younger generation of scientists.

Theses are accepted into the series by invited nomination only and must fulfill all of the following criteria

- They must be written in good English.
- The topic should fall within the confines of Chemistry, Physics, Earth Sciences, Engineering and related interdisciplinary fields such as Materials, Nanoscience, Chemical Engineering, Complex Systems and Biophysics.
- The work reported in the thesis must represent a significant scientific advance.
- If the thesis includes previously published material, permission to reproduce this must be gained from the respective copyright holder.
- They must have been examined and passed during the 12 months prior to nomination.
- Each thesis should include a foreword by the supervisor outlining the significance of its content.
- The theses should have a clearly defined structure including an introduction accessible to scientists not expert in that particular field.

More information about this series at http://www.springer.com/series/8790

Tobias Ostermayr

Relativistically Intense Laser–Microplasma Interactions

Doctoral Thesis accepted by
the Ludwig-Maximilians-Universität München,
Munich, Germany

 Springer

Author
Dr. Tobias Ostermayr
Fakultät für Physik
Ludwig-Maximilians-Universität München
Garching, Germany

Supervisor
Prof. Jörg Schreiber
Fakultät für Physik
Ludwig-Maximilians-Universität München
Garching, Germany

ISSN 2190-5053   ISSN 2190-5061   (electronic)
Springer Theses
ISBN 978-3-030-22210-9   ISBN 978-3-030-22208-6   (eBook)
https://doi.org/10.1007/978-3-030-22208-6

This Springer imprint is published by the registered company Springer Nature Switzerland AG
The registered company address is: Gewerbestrasse 11, 6330 Cham, Switzerland

And a lean, silent figure slowly fades into the gathering darkness, aware at last that in this world, with great power there must also come —great responsibility.
Stan Lee (in Spiderman, Amazing Fantasy #15, Marvel 1962)

Supervisor's Foreword

For more than a century, particle accelerators have represented an important cornerstone for science. Every contemporary photon source for imaging originates with accelerated energetic electrons. Acceleration of the heavy constituents of matter, atomic nuclei, finds application in material analysis, implantation, and radiation therapy, to name a few. Throughout the significant evolution of acceleration technology, conventional techniques continue to rely on the basic principle demonstrated for the first time by Wideroe in 1927. Progress toward achieving higher kinetic energies or better radiation quality has been inevitably connected to the development and availability of radio-frequency power sources. That the laser might now play an important role in advancing particle acceleration is not surprising. Routinely available using chirped pulse amplification, ultrashort laser pulses can now achieve peak powers at the petawatt level. On tightly focusing such a powerful light pulse within the focal volume, target material is almost immediately ionized. Laser fields drive a charge separation as they exert an immense force on the electrons. The much heavier inert atomic ions, largely unaffected by the laser fields, react to the electric fields attributed to this subsequent charge separation. Some laser-plasma-based particle acceleration is closely related to coherent acceleration principles that were postulated by Veksler as early as 1957. However, the situation for implementing laser-driven particle acceleration has been complicated experimentally by a variety of competing processes. Creating practically clear and well-defined laboratory conditions remained elusive before we introduced levitated, fully isolated plasmas. Developed years ago, charged microscopic material can hover in a Paul trap (the basis for the 1989 Nobel Prize for Physics) and these can therefore be used as laser targets. Because the distance from the levitated target to the next proximate material, typically the electrodes of the trap, can be millimeters or more, there is sufficient access for focusing a high-power laser pulse onto the levitated micro-target. As a technical basis for laser-driven acceleration studies, this thesis work describes how the remaining challenge of stably overlapping the micrometer size focus with the micrometer size particle was overcome. This unique solution has provided the basis for unequivocally demonstrating the transition of proton acceleration from the plasma-expansion to the Coulomb-explosion regime.

Interestingly, the exploding heated target does not provide a directed proton beam, which is desired and typically available from conventional accelerators. Instead, it generates a largely divergent spray. In addition, the violently moving electrons not only provide the acceleration fields for the protons but also emit X-rays. This completely new aspect of a combined particle and radiation source was readily exploited in recording a combined proton and X-ray image. As can often occur, some new aspect that appears initially contradictory can also inspire new application capability. The Ph.D. thesis *Relativistically Intense Laser–Microplasma Interactions* represents a formidable example of this instinctively creative wisdom. This research work has demonstrated new technology that can form the basis for high-quality multi-modal imaging sources, further our theoretical understanding and contribute valuable experimental benchmarks for numerical codes. A reading is recommended for those interested in laser-driven plasma acceleration and its unique applications; imaging in particular.

Garching, Germany   Prof. Jörg Schreiber
March 2019

Abstract

This thesis covers several aspects of relativistically intense laser–microplasma interactions. A Paul-trap-based target system was developed that allows to use fully isolated, well-defined and well-positioned micro-sphere-targets in experiments with focused petawatt lasers. The laser–microplasma interaction was then studied in a theoretical framework, in numerical simulations and in a series of experiments. The first key finding of this work was a highly tunable proton beam, well in excess of 10 MeV kinetic energy, that can be modified in spectral and spatial distribution by variations of the acceleration mechanism. This includes the predictable occurrence of **proton beams with a limited spectral bandwidth**, which exceeds similar laser-driven sources in terms of kinetic energy and/or particle fluence. The relative energy bandwidth of down to 20–25% is achieved via the (isotropic) Coulomb explosion or via directional acceleration processes (Fig. 1a, Sects. 5.3 and 5.4). In a second effort, tungsten micro-needle-targets were used at a petawatt laser to produce X-ray and proton beams, both with few-μm effective source-size. While the X-ray spectrum extended up to 10 keV, the proton source again showed the narrow

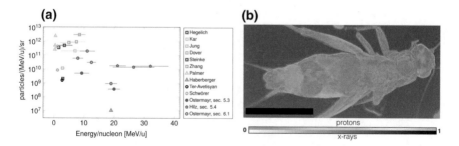

Fig. 1 Key results. a Ion sources with limited spectral bandwidth in this work (red) in the context of earlier experimental (laser-driven) efforts (Refs. [1–10]). Markers are positioned at the spectral peak, error bars represent the FWHM spectral bandwidth. **b** Example of a bi-modal X-ray and proton image (registered on top of each other). The scale bar corresponds to 1 cm. Intensities are given in relative scales and the proton image is shown with 40% transparency

bandwidth (Fig. 1a, Sect. 6.1) with kinetic energies of the peak exceeding 10 MeV. After the full source characterization, single-shot **radiographic imaging** was explored **using the X-rays and the protons simultaneously** (Fig. 1b). The source characteristics also enabled the single-shot recording of X-ray images with phase contrast contribution and the exploration of quantitative single-shot proton radiography.

Publications Related to this Thesis

1. V. Pauw, **T. Ostermayr**, K.-U. Bamberg, P. Böhl, F. Deutschmann, D. Kiefer, C. Klier, N. Moschüring, and H. Ruhl. Particle-in-cell simulation of laser irradiated two-component microspheres in 2 and 3 dimensions. *Nuclear Instruments and Methods in Physics Research Section A (Conference Proceeding EAAC2015)*, 829:372–375, 2016.

2. **T. Ostermayr**, D. Haffa, P. Hilz, V. Pauw, K. Allinger, K.-U. Bamberg, P. Böhl, C. Bömer, P. R. Bolton, F. Deutschmann, T. Ditmire, M. E. Donovan, G. Dyer, E. Gaul, J. Gordon, B. M. Hegelich, D. Kiefer, C. Klier, C. Kreuzer, M. Martinez, E. McCary, A. R. Meadows, N. Moschüring, T. Rösch, H. Ruhl, M. Spinks, C. Wagner, and J. Schreiber. Proton acceleration by irradiation of isolated spheres with an intense laser pulse. *Physical Review E*, 94(3):033208, Sep 2016.

3. **T. Ostermayr**, J. Gebhard, D. Haffa, D. Kiefer, C. Kreuzer, K. Allinger, C. Boemer, J. Braenzel, M. Schnuerer, I. Cermak, J. Schreiber, and P. Hilz. A transportable paul-trap for levitation and accurate positioning of micron-scale particles in vacuum for laser-plasma experiments. *Review of Scientific Instruments*, 89:013302, Jan 2018 (Cover of the January 2018 Issue).

4. P. Hilz, **T. Ostermayr**, A. Huebl, V. Bagnoud, B. Borm, M. Bussmann, M. Gallei, D. Haffa, T. Kluge, F. Lindner, C. Schaefer, U. Schramm, P. Thirolf, T. Rösch, F. Wagner, B. Zielbauer and J. Schreiber, Isolated Proton Bunch Acceleration by a Peta-Watt Laser Pulse, *Nature Communications*, 9:423, Jan 2018.

Acknowledgements

I want to express my gratitude to everyone who helped, joined, and guided me during this adventure.

My first "thank you" goes to Prof. Jörg Schreiber for countless inspiring discussions, great support, and immense trust during all my experiments. I am also very grateful to Peter Hilz and his family Angela, Simon, and Elisabeth. In the past 6 years I've learned a great deal from you. Many thanks to everybody else who was involved with the Paul trap development: Daniel Haffa, Johannes Gebhard, Markus Singer, Ivo Cermak, and Iva Cermakova. The Paul trap, the particle diagnostics, the vacuum chambers, and most other mechanical equipment that was used in the experiments was built by the LMU workshop led by Rolf Oehm, with great advice, expertise, and skill. Designs were supported by the LMU design team led by Dr. Johannes Wulz. I relied on great people during both beamtimes at the Texas Petawatt laser. In my first beamtime, 2014, Daniel Haffa, Daniel Kiefer, Klaus Allinger, and Christina Bömer made it a fun, exciting, and successful time. The 2016 beamtime was another great experience with Christian Kreuzer, Franz Englbrecht, Johannes Gebhard, and Jens Hartmann. The staff at Texas Petawatt, Gilliss Dyer, Mikael Martinez, Michael Donovan, Erhard Gaul, Eddie McCary, Ganesh Tiwari, and Manuel Hegelich, gave us a warm welcome and the possibility to successfully perform our experiments at a unique and top-notch facility. I am grateful to Viktoria Pauw, Karl-Ulrich Bamberg, Hartmut Ruhl, and the PSC team for their work on simulations. Our beamtimes at GSI Darmstadt brought stressful nights and interesting results. The local staff, Bernhard Zielbauer, Florian Wagner, Udo Eisenbarth, Stefan Götte, and Vincent Bagnoud, gave us perfect support. I'm also thankful to our collaborators, Paul Neumayer and Björn Borm, for discussions, contributions, and participation in the experiment. We had great support from Axel Huebl, Thomas Kluge, Michael Bussmann, and Ulrich Schramm for numerical simulations of these experiments. A shoutout goes to the team of the Max-Born-Institute Berlin, especially Matthias Schnürer, Julia Bränzel, Sven Steinke, and Gerd Priebe. They showed me the first TW-class laser I had ever seen, and they gave us the chance to proof our trap concept in late 2012 as complete newcomers. I'm very thankful to Prof. Laszlo Veisz and Daniel Cardenas, who gave me the opportunity to take part in a completely different kind of experiment at their completely different kind of laser system. Thanks to Prof. Katia Parodi, Matthias Würl, and Chiara Gianoli for supporting my work toward quantitative ion radiography in discussion, and for having me at the chair. Thanks to Andrea Leinthaler for great support in all administrative things, to the technological staff at LMU, Uli Friebel, and Simon Stork, and to our fearless driver, Reinhardt Satzkowski. Thank you to all past and current members of the Schreiber group: Florian Lindner, Jens Hartmann, Martin Speicher, Enrico Ridente, Ying Gao, Thomas Rösch, Markus Singer, Sebastian Lehrack, Rong Yang, Jianhui Bin, and Wenjun Ma, you made this group a very enjoyable place to work at. I am very thankful to Professors Toshiki Tajima and Kazuhisha Nakajima, who first taught me about laser-plasma accelerators, and made me very curious. Thanks to my friends Markus Seigfried, Stefan Petrovics, Michael Siegert, and Felix Müller for staying in touch and being good persons. Completing this thesis would have been impossible if it weren't for my family. Thanks to my parents Doris and Richard for livelong support, and to my siblings Dominik and Leonie. Thanks to Christian and Hertha Kremser for being the most friendly human beings I've ever known and thanks to the Schapfls, especially Oma. Thanks also to Leonhard Eberl, Harald Eberl, and Selen for being there during this time. A massive thank you goes to Bine, for spending the past 8 years together with me. You made me stay human!

Contents

Part IV   Summary and Perspectives

Symbols and Abbreviations

c	Speed of light in vacuum (299792458 ms^{-1})
ϵ_0	Vacuum permittivity ($8.854... \times 10^{-12}$ $\text{A}^2\text{s}^4\text{kg}^{-1}\text{m}^{-3}$)
μ_0	Vacuum permeability ($4\pi \times 10^{-7}$ TmA^{-1})
κ_B	Boltzmann constant ($1.38064852(79) \times 10^{-23}$ JK^{-1})
$\hbar$	Planck's constant ($1.054571800(13) \times 10^{-34}$ Js)
e	Elementary charge ($1.6021766208(98) \times 10^{-19}$ C)
e_E	Euler's number ($2.71828...$)
m_e	Electron rest mass ($9.10938356(11) \times 10^{-31}$ kg)
m_p	Proton rest mass ($1.672621898(21) \times 10^{-27}$ kg)
λ	Wavelength (m)
ω, ω_0	Angular frequency (s^{-1})
I	Intensity (Wm^{-2})
E	Energy (J)
eV	Electronvolt (1.60218×10^{-19} J)
$\mathcal{E}$	Electric fields (Vm^{-1})
$\mathcal{E}^*$	Complex conjugated
$\hat{\mathcal{E}}$	Laplace transformed
$\mathcal{B}$	Magnetic fields (T)
T_e, $(T = \kappa_B T_e)$	Temperature (K, (J))
PIC	Particle-in-cell simulation
FWHM	Full width at half maximum
WASP	Ion wide angle spectrometer
TP	Thomson parabola spectrometer
IP	Imaging plate type BAS-TR
CR39	Columbia resin #39 (Tasl UK)
ESF	Edge spread function
LSF	Line spread function
PSF	Point spread function
MTF	Modulation transfer function
WET	Water equivalent thickness

Part I
Introduction and Basics

Chapter 1
Scientific Context and Motivation

In the framework of this thesis, three independent areas of physics-research have been combined to a series of experiments with two central objectives: to increase the understanding of laser-microplasma interactions, and to test their potential for applications by experimental demonstration. The first research-branch relevant to the presented work is the interaction of intense laser pulses with plasmas and the generation of secondary radiation in such. The second area of research are Paul traps that are nowadays a widespread enabling technology for a broad range of physics research as they allow to study matter and its interactions in isolated and controlled states rather than in connected states. The third ingredient to this work is radiographic imaging using high energy particles. In the following sections short overviews of these research areas will be given. The motivation to combine these technologies to new experimental setups will be highlighted in the context of previous efforts.

1.1 Laser Driven Sources of Radiation

Interactions of lasers with matter belong to the most fruitful fields of science ever since their invention in 1960 [1]. The development of laser driven sources of high energy particle radiation started already shortly after the invention of the laser (e.g. [2, 3]). Since then, lasers with ever increasing power, intensity and energy were built (cf. Fig. 1.1). Alongside the advent of laser power, the meaning of 'high energy' in secondary sources of electrons, ions and photons driven by such laser pulses has radically evolved. Thereby, the range of possible applications has been steadily extended [4, 5], with one of the latest additions being medical imaging and therapy.

Early on Q-switched lasers [8] enabled Mega- and Gigawatt lasers with tens to hundreds of Joules energy compressed to ns-range pulse durations. These lasers produced electron and ion kinetic energies of hundred to few kilo-electronvolts by thermally heating the plasma (e.g. [9–12]). Laser plasmas were then amongst

© Springer Nature Switzerland AG 2019
T. Ostermayr, *Relativistically Intense Laser–Microplasma Interactions*,
Springer Theses, https://doi.org/10.1007/978-3-030-22208-6_1

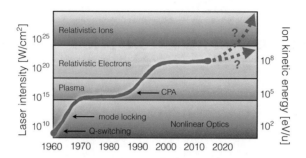

Fig. 1.1 History of laser intensity. The history of laser intensity over the past decades. Adapted from [6, 7]. Especially notable is the slow rise in laser intensity since the early 2000s. This renders engineering and research towards advanced targets and novel laser-ion-acceleration mechanisms particularly important

others considered in context with laser-driven fusion and as injector for magnetically confined laboratory plasmas and particle accelerators. Later, Kerr-lens mode-locking [13–16] brought laser-pulse durations down to several tens to hundreds of fs, with limited pulse energies due to damage thresholds and nonlinearities of optical components in the laser systems.

In 1985 the first demonstration of optical chirped pulse amplification (CPA [17]) for lasers paved the way for today's relativistic-intensity lasers with energies up to ∼100 J delivered in several tens to hundreds of fs pulse duration. When focused onto a target with a spot-size of 1–100 µm, such modern lasers quickly ionize any material, forming a plasma of electrons and positively charged nuclei (ions). During peak-interaction they push electrons close to the speed of light in just about a half laser-cycle or in other words over extremely short distances of half a wavelength (hence the term relativistic). Corresponding accelerating fields (>MV/µm) can exceed fields achieved in conventional electrostatic or radio-frequency accelerators by orders of magnitude, because those are limited by electric-field induced material-breakdown (typically around 0.1–1 kV/µm e.g. [18, 19]). In a plasma, this limitation can be overcome, because it inherently consists of material that has been ionized prior to the peak interaction. Thereby relativistic lasers opened a new realm of particle-acceleration in plasmas, with great potential to shrink required acceleration-lengths. The reduction of accelerator size had been associated with potential cost reduction and therefore considered promising for accelerator-based research and applications. Unlike conventional accelerators, laser-plasmas are sources of a mixed radiation field, whose main contributors are electrons, photons and ions.

Due to their high mobility, electrons are directly acted on by the electro-magnetic fields of the laser-pulse and can gain relativistic velocities during their oscillation over very short distance. Once electrons are relativistic, they can be be very efficiently accelerated further in co-moving structures, e.g. created by the laser propagating through an underdense plasma. Such laser-wakefield (a plasma wave of electron density modulations on top of the quasi-unperturbed ionic background) creates electric fields that can be 'surfed' by electrons to be both focused and accelerated when injected in the suitable phase. Laser-wakefield acceleration of electrons in underdense plasmas [20] has achieved significant results as table top-accelerator generating useful monochromatic electron-beams [21]. Recently, first steps addressing scalability of laser-wakefield acceleration towards large-scale accelerators and collider-facilities have been taken [22].

Whenever electrons are being accelerated and heated by the extreme laser intensity, they emit energetic photon radiation, which itself can be optimized and utilized. Of particular interest is the possibility to confine the temporal and spatial extent of the source for fast imaging applications while maintaining a significantly high photon number per single pulse. Modern fourth generation light-sources driven by conventional accelerators set standards in many respects (e.g. [23]), resolving nm and fs scales and providing transverse coherence lengths in the mm range, but requiring large scale facilities and long imaging beamlines. Laser driven table top sources as demonstrated in Refs. [24–32] and in the framework of this thesis explore another means to produce applicable photon sources, making use of the intrinsic spatio-temporal source limitation dictated by the high-power laser pulse. Despite being laser-driven, there are several distinct methods to produce useful X-rays in a wide spectral range. This includes betatron radiation [24–26], Thomson backscattering [27], high harmonic generation [32], Bremsstrahlung, recombination and line emission [33, and references therein]. In fact, even the generation of the shortest bursts of light (several attoseconds long, in the UV spectral range) relies on laser pulses with comparably moderate intensities (i.e. non-relativistic) interacting with gaseous targets, exploiting the coherent nonlinear motion of electrons in the electric field of the laser [34].

Similar to the generation of X-rays via accelerated electrons, energetic ion-beams from laser-plasma interactions will be produced only indirectly with today's lasers, since coupling to the much heavier ions is insufficient to drive significant direct ion-acceleration. For example, with a laser intensity of 10^{20} Wcm^{-2} and a wavelength of $1\,\mu$m, as used in this work, the oscillation of a proton in the laser field occurs at velocities of only 0.5% of the speed of light. However, it has been established that quasi static charge-separation fields can emerge when electrons are heated to large (MeV) temperatures in a solid density target that partly reflects and partly absorbs the laser pulse, leading to an expansion of the plasma into vacuum. Such non-oscillating charge-separation fields can be as large as the laser-field itself (MV/μm) and persist for long enough to indirectly accelerate ions to significant and potentially useful kinetic energies [35–37]. An example for the typical ion energy distribution from laser-driven ion sources is shown in Fig. 1.2a; it shows the characteristic broad bandwidth, with decreasing particle numbers towards higher energies, and with a defined cutoff energy. Although cutoff energies beyond 85 MeV have been reported [38–40], the spectral shape is problematic for many applications. Many efforts have been undertaken since the 2000s to produce mono-energetic (or bandwidth-limited so-called quasi mono-energetic) ion pulses, which are summarized in Fig. 1.2b. The gray shaded area highlights the largest issue with quasi mono-energetic sources to date: either the energy or the particle count (or both) are very limited (especially when compared to non-monoenergetic sources).

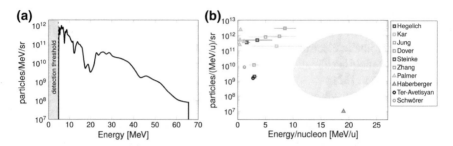

Fig. 1.2 Typical proton spectrum and summary of efforts showing quasi-monoenergetic ions.
a A typical ion kinetic energy distribution measured in a laser-plasma interaction of the Texas
Petawatt laser with a 200 nm thin polymer foil (recorded in one of our experiments). **b** Summary
of experiments showing quasi-monoenergetic ion kinetic energy distributions enabled by different
manipulations of targets and/or lasers. Squares show foil-based experiments, triangles show near-
critical plasmas, circles show mass-limited targets. The peak ion count and the full width at half
maximum (error bars) are specified. References: Hegelich [41], Kar [42], Jung [43], Dover [44],
Steinke [45], Zhang [46], Palmer [47], Haberberger [48], Ter-Averisyan [49], Schwoerer [50]

1.2 Paul Traps and Isolated Targets in Laser-Plasma Interactions

Physicists aim to understand the structure of matter and the forces connecting it.[1] A
natural problem for the precise observation of matter is, that it is typically connected
to more matter, making any measurement an average over an ensemble of particles.
The wish to study isolated matter in order to gain information on its isolated behavior
triggered the development of particle traps. During the 1950s it was recognized in
molecular beam physics [52–54], *mass spectrometry* [55–57] *and particle acceler-*
ator physics that plane electric and magnetic multipole fields can *focus* particles *in*
two dimensions [51]. This was amongst others used in the development of masers
[58].

Linear quadrupole mass spectrometers use the *focusing and defocusing properties*
of the high frequency electric quadrupole field together with the *stability properties*
of their equations of motion [51] to keep particles of appropriate charge-to-mass ratio
close to the spectrometer axis. Extension of these principles to the third dimension
enabled the construction of particle traps [59–61]. The possibility to measure individual particles opened new dimensions in atomic physics, mass spectrometry, storage
and research of antimatter, optical and microwave-spectroscopy, quantum effects,
quantum information processing, and molecular physics to name but a few fields
that are still very active today (e.g. [62], and references therein). Particle traps also
play a vital role in investigations of non-neutral plasmas [63, 64], usually formed
by large clouds of trapped charged particles. Experiments presented herein could
be regarded as a very distant relative of such experiments in a high-density and
high-energy limit.

---

[1]The first two paragraphs here are based on the story-line in Ref. [51].

Similar to other fields of research, the understanding of laser driven sources in many cases suffers from the difficult interpretation of experimental results. Parts of this difficulty can be attributed to the extent of non-isolated targets. The typical spatial laser intensity distribution covers 10^{12}–10^{21} Wcm^{-2} over $\sim$100 μm. Thus, a large number of particles originating from the diverse physical processes occurring at different intensity levels usually contribute to the (integrated) measured signal.

Already early after the invention of the laser, the use of isolated targets in laser-matter interaction has been suggested and tested [9–12]. The first motivation to use isolated targets in laser plasma experiments is therefore nothing different from the motivation for using isolated objects in other fields of research; the limitation of electron/ion number in the target and the limitation of the laser-target cross-section enable a more detailed understanding of the laser plasma interaction by reducing unknown parameters. This is not only relevant for laser-driven particle sources, but more generally for studies that rely on the laser-plasma interaction, including studies of warm dense matter (WDM) [65–67].

Another motivation emerges from a more practical point of view. Some applications of laser driven sources will greatly benefit from a defined number of accelerated particles originating from a well-known source with controlled size in the sub-micron range, i.e. much smaller than typically achieved even with laser-driven sources using bulk foil-targets. *This includes proton- and X-ray-imaging [24, 68], laser-absorption measurements at ultra-high intensities via spectroscopy of all particles contained in a known target (complementary to light-measurements [69]) and could even range to laboratory-scale astrophysics [70]* (originally published in Ref. [71]). As an additional motivation, the isolation of target particles has often been suggested in calculations and numerical simulations as an efficient means to positively influence ion kinetic energies and to monochromatize ion-energy distributions from laser driven sources via several mechanisms (e.g. [72–79]) and for generation of ultrashort relativistic electron bunches [80, 81].

In the past decade, so-called mass limited targets have been approached using targets in the few 10-μm range held in position by spider-silk [82] or other tiny mounts in the 100 nm to few μm range [65–67, 83, 84], isolated water-droplets with diameters typically in the 10 μm range [85–87] and gas-cluster targets in the 100 nm range [49, 88, 89]. In all these examples, target diameters accessible via one single technique were very limited. Futhermore, the next solid structure (e.g. the next target) is directly connected or nearby the target, at distances of only few micrometers. Additionally, targets from droplet and cluster sources are often surrounded by significant amounts of gas. Generally, nearby material is known to contribute to the interaction [49, 83, 85] (originally published in Ref. [71]) *even if its mass and scale are tiny compared to the actual target. This has important implications, even for large scale (mm) plasmas used in inertial confinement fusion [90].*

The typical laser-focus spot diameter in such experiments is in the 1–10 μm range and current PW lasers have limited repetition rates. For these reasons, investigations of single shot laser-target interactions with targets of the size of the focus itself or even smaller demand for highest accuracy in target and focus positioning in order to obtain meaningful results. It appears, that suspending the target in an electrodynamic trap is the natural solution to solve this problem with a fully isolated target that comes without *any* strings attached.

1.3   Radiographic Imaging

Radiographic imaging using high energy photons for medical purpose was first enabled by Röntgen's discovery of X-radiation in 1895 [91–93]. The radiographic contrast in an X-ray radiography is usually generated by different attenuation cross sections for X-rays in different kinds of tissue. Since then, much effort has been spent to reduce the dose that is deposited in imaged patients, exposure times, as well as to improve image quality. First computer-tomographs (CT) were built in 1969 and commercialized in 1972 by Hounsfield [94], enabling three dimensional imaging of patients. Modern CT use high-end robotics and computational techniques to reconstruct high resolution 3D information from a series of images [95]. Together with magnetic resonance imaging [96, 97], which relies on the nuclear spin of protons in the imaged sample instead of highly energetic photons interacting with the electrons in atoms, CT is the standard of modern biomedical imaging.

More recently, novel advances in X-ray source and detector developments enabled studies that—instead of the X-ray attenuation image—allow to generate radiographic contrast using the phase-shift a sample imprints on the X-ray beam, via quasi interferometric methods (e.g. [98–101]). This so-called phase-contrast imaging (PCI) can provide enhanced contrast for features that are difficult to recognize in conventional attenuation images, e.g. due to comparable attenuation of an object and its surrounding material.

Clearly, X-rays have revolutionized medicine, giving unprecedented views into objects and living humans for medical diagnostics. Current efforts at fourth generation light sources [23] use X-rays for studies in material science, warm dense matter and many more fields of research using ultrahigh spatio-temporal (nm, fs) resolution.

Alongside the long-term development of X-ray imaging, much effort has been spent to produce images by use of proton stopping (e.g. [102–105]). Radiographies and tomographies using proton beams from conventional accelerators were demonstrated in close context with charged particle therapy of cancer. When ions penetrate into matter, most of their energy is deposited near the Bragg peak close to the maximum range of the ions in the material, as depicted in Fig. 1.3. In contrast, the depth-dose curve of MeV scale photons decays exponentially after the initial buildup (lower energy photons do not exhibit the build-up). In principle that makes it much easier for protons than for photons to concentrate the dose on a tumor, while simultaneously reducing the dose on healthy tissue (and vital organs) surrounding it. However, to date the largest uncertainty in charged particle therapy of cancer (i.e. therapy with ions) originates from treatment-planning based on X-ray CT while irradiating the tumor with protons [106].

Proton imaging holds the potential to deliver a more direct and accurate method of image acquisition, circumventing the difficile conversion from X-ray attenuation

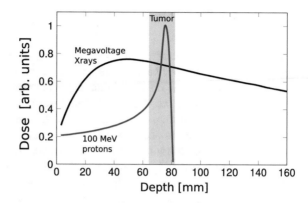

Fig. 1.3 Bragg peak. Depth dose for X-rays and protons

to proton stopping, which is used to plan the therapeutic irradiation. In addition, the dose deposited in a patient during the scan can be reduced by a factor of 10–100 as compared to X-ray CT, due to the smaller amount of integrated energy loss of ions along their path, when penetrating completely through the patient [107, 108] (i.e. adjusting the energy such that the Bragg peak is located behind the patient). On the other hand, the production of useful proton images is challenging and costly. The first reason is, that conventional proton accelerators are much larger machines than X-ray sources, and can involve ion-beam imaging optics that contribute to the overall size [105]. Second, without dedicated efforts, images appear blurred due to the statistical nature of multiple Coulomb scattering (MCS) and range straggling, and thereby suffer from limited resolution.

All of today's clinical radiographic imaging relies on conventional accelerator technology for generating the radiation. As such they are mono-species sources of mostly monoenergetic radiation. Imaging experiments using intense laser-driven sources of energetic particles have thus far aimed to asses their applicability in many scenarios close to ones known from conventional accelerators. This includes the focus on a single species of radiation, disregarding respective other kinds of energetic particles originating from the laser-plasma interaction (as mentioned earlier) as a mere disturbance, which needs to be suppressed (difficult in a laser-plasma) or eliminated from the measured signal. The same is true for experiments using laser-driven photon or proton sources as probes for high-energy-density matter and laser-plasma interactions [29, 30, 68, 109–113].

Early work with laser driven sources tried to exploit the multi-modal emission to record simultaneous X-ray and proton images, but was limited to binary imaging of thin mesh objects [114, 115]. The approach faced some fundamental problems, which had to be resolved before imaging thicker (more relevant) objects in more than a 'black-and-white' mode would become feasible, as we will discuss in this thesis.

1.4   Aim of This Work

The first major objective of this thesis is the development, characterization and implementation of a Paul trap based target system to provide truly isolated nano- and
micro-targets of various material in an unprecedented range of target size at high
power (TW-PW) laser facilities, to explore laser interactions with a confined and
well-known micro-plasma. An earlier approach using a Paul trap was quite limited in terms of position accuracy and range of target mass, due to rather simple
techniques used for charging particles and stabilizing their trajectory [116]. Our
newly developed system overcomes these limitations by improving the operation. It
implements advanced techniques including ion-beam charging and active instead of
passive trajectory damping. The system gives the unique opportunity to controllably
study isolated micro-plasmas with sharp boundaries at the target-vacuum interface
as well as purposefully pre-expanded targets. It was successfully implemented in a
test experiment at the Max Born Institute Berlin (MBI) that will not be discussed
here, and in complementary experiments at the **T**exas **P**eta-**W**att laser (TPW) at the
University of Texas at Austin and at the **P**etawatt **H**igh-**E**nergy **L**aser for **H**eavy **I**on
EXperiments (PHELIX) at the Gesellschaft für Schwerionenforschung, Darmstadt.

In this first series of laser-plasma experiments, we investigate the dynamics of
laser driven micro-sources of ions. The detailed knowledge of the laser and target
conditions enables uniquely clean insights to ion acceleration processes in such
kind of system, and gives access to hitherto unavailable parameters in quasi monoenergetic ion sources in terms of energy and particle count (gray region in Fig. 1.2b).

In a second experimental effort, we use tungsten nano-needle targets at the Texas
Petawatt laser. After characterizing the limited bandwidth and the effective size of the
source, we make use of its multi-modal radiation field, to simultaneously record high-
resolution and high-contrast X-ray and proton images in a single laser-shot. With this
target, the transmitted laser and high energy electron beams are emitted primarily
along the unaltered laser propagation direction. Meanwhile, protons are emitted
around the entire plane normal to the needle and X-rays are emitted isotropically.
This separation of protons and X-rays from the unwanted laser and electron beams
allowed to record images of both modalities without contamination by other radiation.
Due to the small size of the initial target, source size remains small despite the larger
amount of laser energy compared to early work. Proton energies up to 20 MeV and
few-keV X-rays allow imaging of technological and biological samples in the mm-
cm scale. As we will discuss, the quasi-monoenergetic proton distribution facilitates
high contrast imaging in a mode that clearly goes beyond binary imaging and may
finally facilitate multi-mode applications.

Our experiments elucidate possible benefits of future laser-driven ion-accelerators
when considering their application in the medical sector, making optimum use of the
novel accelerator's properties (e.g. its finite bandwidth and multi-modal radiation
field) instead of just trying to duplicate existing conventional radiation sources (e.g.
[117]). The optimum timing between all sources due to the common drive laser,

and its comparably simple implementation, also make it an interesting candidate for research applications in ultrafast high-energy-density science.

1.5 Thesis Structure

Chapter 2 discusses theoretical basics of laser-plasmas that are relevant to all topics touched in this work. Details relevant only to sub-topics of this thesis will be given in the corresponding chapters.

Chapter 3 addresses experimental methods of high power lasers, and specific problems that occur when using them with isolated targets.

Chapter 4 introduces the portable Paul trap system, developed to provide and position fully isolated micro-targets in vacuum.

Chapter 5 presents theoretical expectations, numerical simulations and experiments performed with levitating spherical targets, with a dedicated focus on ion acceleration.

Chapter 6 presents experiments investigating radiographic imaging using a laser-driven micro-source. Simultaneous proton-imaging and X-ray imaging (including phase-contrast imaging) from a single source is explored using the unique properties of laser-driven micro-plasma sources.

Chapter 7 gives a summary of presented work and central results.

Chapter 8 delivers an outlook on novel opportunities that are based on the achievements presented herein.

Note on laser beam-times This note is meant to clarify the special nature of experimental environments that were used to record data presented in this thesis. The Texas Petawatt of UT Austin as well as the PHELIX laser at GSI Darmstadt are large-scale facilities that require significant amounts of space, infrastructure, maintenance and full-time staff for operation. Both systems have a limited number of laser beam-time-slots per year available for external scientists (such as us). In order to receive one of these precious time-slots, detailed experiment proposals need to be submitted to expert committees, who distribute beam-times according to proposal quality, impact, likelihood of success and more criteria. While detailed statistics are not officially provided, it is well known for both lasers, that the number of proposals regularly outnumbers the number of available time-slots per year.

A laser beam-time at such large-scale facility is typically limited in duration to several weeks. Before an experiment, the experiment leader prepares the detailed plan of scientific objectives, setup plans, logistics of required material (e.g. we shipped equipment worth 500 kEUR to UT before our first run, and back to Munich afterwards), laser-shot plans and specific tasks for each team-member. During the setup-time and experimental run, a team consisting of up to 6 external scientists supported by local staff-members assembles the experiment and operates relevant equipment and diagnostics, according to the detailed plans of the experiment leader. The laser itself is operated by dedicated local staff. In such complex experiments that are comprised of a number of sub-experiments, the principal scientist usually assigns a deputy to deal with a subset of tasks and team-members. Since any experiment of this kind involves more than the effort of a single scientist, the specific contribution of the author to experiments presented herein will be specified in footnotes accompanying the respective chapter.

References

1. Maiman TH (1960) Stimulated optical radiation in ruby. Nature 187:493–494
2. Honig RE, Woolston JR (1963) Laser-induced emission of electrons, ions, and neutral atoms from solid surfaces. Appl Phys Lett 2(7):138–139
3. Giori F, MacKenzie LA, McKinney EJ (1963) Laser-induced thermionic emission. Appl Phys Lett 3(2):25–27
4. Daido H, Nishiuchi M, Pirozhkov AS (2012) Review of laser-driven ion sources and their applications. Rep Prog Phys 75(5):056401
5. Esarey E, Schroeder CB, Leemans WP (2009) Physics of laser-driven plasma-based electron accelerators. Rev Mod Phys 81(3):1229–1285
6. Schreiber J (2006) Ion acceleration driven by high-intensity laser pulses. PhD thesis, Ludwig-Maximilians-Universität, München
7. Mourou GA, Tajima T, Bulanov SV (2006) Optics in the relativistic regime. Rev Mod Phys 78(2):309–371
8. McClung FJ, Hellwarth RW (1962) Giant optical pulsations from ruby. J Appl Phys 33(3):828–829
9. Dawson JM (1964) On the production of plasma by giant pulse lasers. Phys Fluids 7:981
10. Haught AF, Polk DH (1966) High-temperature plasmas produced by laser beam irradiation of single solid particles. Phys Fluids 9(10):2047
11. Jarboe TR (1974) Study of an isolated, laser-produced deuterium plasma in a magnetic field. Dissertation. Lawrence Berkeley Laboratory and UC Berkeley
12. Baumhacker H et al (1977) Plasma production by irradiating freely falling deuterium pellets with a highpower laser. Appl Phys Lett 30(9):461–463
13. Brabec T et al (1992) Kerr lens mode locking. Opt Lett 17(18):1292–1294
14. Kerr J (1875) LIV. A new relation between electricity and light: dielectrified media birefringent. Philos Mag Ser 4 50(332):337–348
15. Kerr J (1875) LIV. A new relation between electricity and light: dielectrified media birefringent (Second paper). Philos Mag Ser 4 50(333):446–458
16. Spence DE, Kean PN, Sibbett W (1991) 60-fsec pulse generation from a self-mode-locked Ti:sapphire laser. Opt Lett 16(1):42–44
17. Strickland D, Mourou G (1985) Compression of amplified chirped optical pulses. Opt Commun 56(3):219–221
18. Matsumoto S et al (2011) High gradient test at Nextef and high-power longterm operation of devices. Nucl Instrum Methods Phys Res Sect A Accel Spectrom Detect Assoc Equip 657(1):160–167
19. Degiovanni A, Wuensch W, Navarro JG (2016) Comparison of the conditioning of high gradient accelerating structures. Phys Rev Accel Beams 19:032001
20. Tajima T, Dawson JM (1979) Laser electron accelerator. Phys Rev Lett 43:267–270
21. Leemans WP et al (2006) GeV electron beams from a centimetre-scale accelerator. Nat Phys 2(10):696–699
22. Steinke S et al (2016) Multistage coupling of independent laser-plasma accelerators. Nature 530(7589):190–193

23. Bostedt C et al (2013) Ultra-fast and ultra-intense x-ray sciences: first results from the linac coherent light source free-electron laser. J Phys B At Mol Opt Phys 46(16):164003
24. Kneip S et al (2010) Bright spatially coherent synchrotron X-rays from a table-top source. Nat Phys 6(12):980–983
25. Kneip S et al (2011) X-ray phase contrast imaging of biological specimens with femtosecond pulses of betatron radiation from a compact laser plasma wake-field accelerator. Appl Phys Lett 99(9):093701
26. Wenz J et al (2015) Quantitative X-ray phase-contrast microtomography from a compact laser-driven betatron source. Nat Commun 6
27. Khrennikov K et al (2015) Tunable all-optical quasimonochromatic thomson x-ray source in the nonlinear regime. Phys Rev Lett 114(19):195003
28. Faenov AY et al (2015) Nonlinear increase of X-ray intensities from thin foils irradiated with a 200 TW femtosecond laser. Sci Rep 5:13436
29. Barrios MA et al (2014) X-ray area backlighter development at the national ignition facility (invited). Rev Sci Instrum 85(11):11D502
30. Jarrott LC et al (2014) K and bremsstrahlung x-ray radiation backlighter sources from short pulse laser driven silver targets as a function of laser pre-pulse energy. Phys Plasmas 21(3):031211
31. Weisshaupt J et al (2014) High-brightness table-top hard X-ray source driven by sub-100-femtosecond mid-infrared pulses. Nat Photonics 8(12):927–930
32. Dromey B et al (2006) High harmonic generation in the relativistic limit. Nat Phys 2(7):456–459
33. Pfeifer T, Spielmann C, Gerber G (2006) Femtosecond x-ray science. Rep Prog Phys 69(2):443–505
34. Krausz F, Ivanov M (2009) Attosecond physics. Rev Mod Phys 81(1):163–234
35. Snavely RA et al (2000) Intense high energy proton beams from Petawatt-laser irradiation of solids. Phys Rev Lett 85(14)
36. Wilks SC et al (2001) Energetic proton generation in ultra-intense laser-solid interactions. Phys Plasmas 8(2):542–549
37. Hatchett SP et al (2000) Electron, photon, and ion beams from the relativistic interaction of Petawatt laser pulses with solid targets. Phys Plasmas 7(5):2076
38. Wagner F et al (2016) Maximum proton energy above 85 MeV from the relativistic interaction of laser pulses with micrometer thick CH2 targets. Phys Rev Lett 116(20):205002
39. Kim IJ et al (2016) Radiation pressure acceleration of protons to 93MeV with circularly polarized Petawatt laser pulses. Phys Plasmas 23(7):070701
40. Hegelich BM et al (2013) 160 MeV laser-accelerated protons from CH2 nanotargets for proton cancer therapy. ArXiv e-prints
41. Hegelich BM et al (2006) Laser acceleration of quasi-monoenergetic MeV ion beams. Nature 439:441
42. Kar S et al (2012) Ion acceleration in multispecies targets driven by intense laser radiation pressure. Phys Rev Lett 109(18)
43. Jung D et al (201) Monoenergetic ion beam generation by driving ion solitary waves with circularly polarized laser light. Phys Rev Lett 107:115002
44. Dover NP et al (2016) Buffered high charge spectrally-peaked proton beams in the relativistic-transparency regime. New J Phys 18(1):013038
45. Steinke S et al (2013) Stable laser-ion acceleration in the light sail regime. Phys Rev Spec Top Accel Beams 16(1):011303
46. Zhang H et al (2017) Collisionless shock acceleration of high-flux quasimonoenergetic proton beams driven by circularly polarized laser pulses. Phys Rev Lett 119(16)
47. Palmer CAJ et al (2011) Monoenergetic proton beams accelerated by a radiation pressure driven shock. Phys Rev Lett 106(1):014801
48. Haberberger D et al (2012) Collisionless shocks in laser-produced plasma generate monoenergetic high-energy proton beams. Nat Phys 8:95–99

49. Ter-Avetisyan S et al (2012) Generation of a quasi-monoergetic proton beam from laser-irradiated sub-micron droplets. Phys Plasma 19(7):073112
50. Schwoerer H et al (2006) Laser-plasma acceleration of quasi-monoenergetic protons from microstructured targets. Nature 439(7075):445–448
51. Paul W (1993) Electromagnetic traps for charged and neutral particles. In: Frä ngsmyr T, Ekspong G (eds) Nobel Lectures, Physics 1981–1990. World Scientific Publishing Co., Singapore
52. Friedburg H, Paul W (1951) Optische Abbildungen mit neutralen Atomen. Naturwissenschaften 38(7):159
53. Bennewitz HG, Paul W (1954) Eine Methode zur Bestimmung von Kernmomenten mit fokussiertem Atomstrahl. Zeitschrift für Physik A Hadrons and Nuclei 139(5):489–497
54. Bennewitz HG, Paul W, Schlier Ch (1955) Fokussierung polarer moleküle. Zeitschrift für Physik A Hadrons and Nuclei 141(1):6–15
55. Paul W, Steinwedel H (1953) Ein neues Massenspektrometer ohne Magnetfeld. Zeitschrift für Naturforschung A 8:448–450
56. Paul W, Raether, M (1955) Das elektrische massenfilter. Zeitschrift für Physik 140(3):262–273
57. Dawson PH (2013) Quadrupole mass spectrometry and its applications. Elsevier Science
58. Townes CH (1983) Science, technology, and invention: their progress and interactions. Proc Natl Acad Sci 80(24):7679–7683
59. Paul W, Osberghaus O, Fischer E (1958) Ein Ionenkäfig. de. Wiesbaden: VS Verlag für Sozialwissenschaften
60. Fischer E (1959) Die dreidimensionale Stabilisierung von Ladungsträgern in einem Vierpolfeld. Zeitschrift für Physik 156(1):1–26
61. Paul W (1990) Electromagnetic traps for charged and neutral particles. Rev Mod Phys 62(3):531
62. Knoop M, Madsen N, Thompson RC (eds) (2014) Physics with trapped charged particles. Imperial College Press, London
63. Bollinger J, Wineland DJ, Dubin DHE (1994) Non-neutral ion plasmas and crystals, laser cooling, and atomic clocks. Phys Plasmas 1(5):1403–1414
64. Werth G (2005) Non-neutral plasmas and collective phenomena in ion traps. In: Dinklage A, et al (eds) Plasma physics: confinement, transport and collective effects. Springer, Berlin, Heidelberg, pp 269–295
65. Myatt J et al (2007) High-intensity laser interactions with mass-limited solid targets and implications for fast-ignition experiments on OMEGA EP. Phys Plasmas 14(5):056301
66. Nishimura H et al (2011) Energy transport and isochoric heating of a low-Z, reduced-mass target irradiated with a high intensity laser pulse. Phys Plasmas 18(2):022702
67. Neumayer P et al (2009) Isochoric heating of reduced mass targets by ultra-intense laser produced relativistic electrons. High Energy Density Phys 5(4):244–248
68. Rygg JR et al (2008) Proton radiography of inertial fusion implosions. Science 319(5867):1223–1225
69. Ping Y et al (2008) Absorption of short laser pulses on solid targets in the ultrarelativistic regime. Phys Rev Lett 100(8):085004
70. Albertazzi B et al (2014) Laboratory formation of a scaled protostellar jet by coaligned poloidal magneticfield. Science 346(6207):325–328
71. Ostermayr TM et al (2016) Proton acceleration by irradiation of isolated spheres with an intense laser pulse. Phys Rev E 94(3):033208
72. Yu W et al (2005) Direct acceleration of solid-density plasma bunch by ultraintense laser. Phys Rev E 72(4):046401
73. Bulanov SS et al (2008) Accelerating monoenergetic protons from ultrathin foils by at-top laser pulses in the directed-Coulomb-explosion regime. Phys Rev E 78(2):026412
74. Kluge T et al (2010) Enhanced laser ion acceleration from mass-limited foils. Phys Plasmas 17(12):123103
75. Limpouch J et al (2008) Enhanced laser ion acceleration from mass-limited targets. Laser Part Beams 26(02):225–234

76. Peano F, Fonseca RA, Silva LO (2005) Dynamics and control of shock shells in the coulomb explosion of very large deuterium clusters. Phys Rev Lett 94(3):033401
77. Peano F et al (2006) Kinetics of the collisionless expansion of spherical nanoplasmas. Phys Rev Lett 96(17):175002
78. Murakami M, Tanaka M (2008) Nanocluster explosions and quasimonoenergetic spectra by homogeneously distributed impurity ions. Phys Plasmas 15:082702
79. Di Lucchio L, Andreev AA, Gibbon P (2015) Ion acceleration by intense, few-cycle laser pulses with nanodroplets. Phys Plasmas 22(5)053114
80. Di Lucchio L, Gibbon P (2015) Relativistic attosecond electron bunch emission from few-cycle laser irradiated nanoscale droplets. Phys Rev ST Accel Beams 18(2):
81. Liseykina TV, Pirner S, Bauer D (2010) Relativistic attosecond electron bunches from laser-illuminated droplets. Phys Rev Lett 104(9):095002
82. Brinker BA et al (1983) Inertial fusion target mounting methods: new fabrication procedures reduce the mounting support perturbation. J Vac Sci Technol A Vac Surf Films 1(2):941–944
83. Henig A et al (2009) Laser-driven shock acceleration of ion beams from spherical mass-limited targets. Phys Rev Lett 102(9):095002
84. Zeil K et al (2014) Robust energy enhancement of ultrashort pulse laser accelerated protons from reduced mass targets. Plasma Phys Control Fusion 56:084004
85. Sokollik T et al (2009) Directional laser-driven ion acceleration from microspheres. Phys Rev Lett 103:135003
86. Karsch S et al (2003) High-intensity laser induced ion acceleration from heavy- water droplets. Phys Rev Lett 91(1):015001
87. Ter-Avetisyan S et al (2006) Quasimonoenergetic deuteron bursts produced by ultraintense laser pulses. Phys Rev Lett 96(14):145006
88. Ditmire T et al (1999) Fusion from explosions of femtosecond laser-heated nuclear fusion from explosions of femtosecond laser-heated deuterium clusters. Nature 398:489
89. Fukuda Y et al (2009) Energy increase in multi-MeV ion acceleration in the interaction of a short pulse laser with a cluster-gas target. Phys Rev Lett 103(16):165002
90. Nagel SR (2009) Studies of electron acceleration mechanisms in relativistic laser-plasma interactions. PhD thesis, Imperial College London
91. Röntgen WC (1895) Ueber eine neue Art von Strahlen. Vorläufige Mitteilung. In: Sitzungs-berichte der Physikalisch-Medicinsichen Gesellschaft zu Würzburg 9. Würzburg: Verlag und Druck der Stahel'schen K. Hof- und Universitäts- Buch- und Kunsthandlung, pp 132–141
92. Röntgen WC (1896) Ueber eine neue Art von Strahlen. 2. Mitteilung. In: Sitzungsberichte der Physikalisch-Medicinsichen Gesellschaft zu WPürzburg. Würzburg: Verlag und Druck der Stahel'schen K. Hof- und Universitäts-Buch- und Kunsthandlung, pp 1–9
93. Röntgen WC (1897) Weitere Beobachtungen über die Eigenschaften der X-Strahlen. In: vol. Band Erster Halbband. Sitzungsberichte der Königlich Preußischen Akademie der Wis-senschaften zu Berlin. Berlin: Verl. d. Kgl. Akad. d. Wiss., pp 576–592
94. Hounsfield GN (1973) Computerized transverse axial scanning (tomography): part 1. Descrip-tion of system. Br J Radiol 46(552):1016–1022
95. Buzug TM (2008) Computed tomography—From photon statistics to modern cone-beam CT. Springer, Berlin Heidelberg
96. Odeblad E, Lindström G (1955) Some preliminary observations on the proton magnetic res-onance in biological samples. Acta Radiol 43:469–476
97. Lauterbur PC (1973) Image formation by induced local interactions: examples of employing nuclear magnetic resonance. Nature 242(5394):190–191
98. Davis TJ et al (1995) Phase-contrast imaging of weakly absorbing materials using hard X-rays. Nature 373(6515):595–598
99. Ingal VN, Beliaevskaya EA (1995) X-ray plane-wave topography observation of the phase contrast from a non-crystalline object. J Phys D Appl Phys 28(11):2314
100. Wilkins SW et al (1996) Phase-contrast imaging using polychromatic hard X rays. Nature 384(6607):335–338

101. Pfeiffer F et al (2006) Phase retrieval and differential phase-contrast imaging with low-brilliance X-ray sources. Nat Phys 2(4):258–261
102. Steward VW, Koehler AM (1973) Proton radiographic detection of strokes. Nature 245(5419):38–40
103. Steward VW, Koehler AM (1973) Proton beam radiography in tumor detection. Science 179(4076):913–914
104. Koehler AM (1968) Proton Radiogr. Science 160(3825):303–304
105. Prall M et al (2016) High-energy proton imaging for biomedical applications. Sci Rep 6:27651 EP
106. Hansen DC (2014) Improving ion computed tomography. PhD thesis, Aarhus University
107. Hanson KM et al (1982) Proton computed tomography of human specimens. Phys Med Biol 27(1):25
108. Hernández LM (2017) Low-dose ion-based transmission radiography and tomography for optimization of carbon ion-beam therapy. PhD thesis, Ludwig- Maximilians-Universität München
109. Borghesi M et al (2001) Proton imaging: a diagnostic for inertial confinement fusion/fast ignitor studies. Plasma Phys Control Fusion 43(12A):A267–A276
110. Sokollik T et al (2008) Transient electric fields in laser plasmas observed by proton streak deflectometry. Appl Phys Lett 92(9):091503
111. Park H-S et al (2008) High-resolution 17–75keV backlighters for high energy density experiments. Phys Plasmas 15(7):072705
112. Brambrink E et al (2009) Direct density measurement of shock-compressed iron using hard x rays generated by a short laser pulse. Phys Rev E 80(5):
113. Brambrink E et al (2009) X-ray source studies for radiography of dense matter. Phys Plasmas 16(3):033101
114. Orimo S et al (2007) Simultaneous proton and x-ray imaging with femtosecond intense laser driven plasma source. Jpn J Appl Phys 46(9A):5853–5858
115. Nishiuchi M et al (2008) Laser-driven proton sources and their applications: femtosecond intense laser plasma driven simultaneous proton and x-ray imaging. J Phys Conf Ser 112:042036
116. Sokollik T et al (2010) Laser-driven ion acceleration using isolated mass-limited spheres. New J Phys 12(11):113013
117. Hofmann KM (2015) Feasibility and optimization of compact laser-driven beam lines for proton therapy: a treatment planning study. PhD thesis, Technische Universität München

Chapter 2
Laser-Plasmas

This chapter introduces fundamental concepts of light, plasma and laser-plasma-interactions, focusing on realms relevant to this work. Wherever necessary, the basic framework is extended later in this thesis, within the corresponding chapters.

2.1 Electromagnetic Waves

Electromagnetic fields can be fully described in the framework of Maxwell's equations[1]

$$\nabla \vec{\mathcal{E}} = \frac{\rho}{\epsilon_0}, \tag{2.1}$$

$$\nabla \vec{B} = 0, \tag{2.2}$$

$$\nabla \times \vec{\mathcal{E}} = -\frac{\partial \vec{B}}{\partial t}, \tag{2.3}$$

$$\nabla \times \vec{B} = \mu_0 \left(\vec{j} + \epsilon_0 \frac{\partial \vec{\mathcal{E}}}{\partial t} \right). \tag{2.4}$$

$\vec{\mathcal{E}}$ and $\vec{B}$ are the electric field and magnetic field respectively, ρ is the charge density, $\vec{j}$ denotes the current density, and ϵ_0 and μ_0 are the permittivity and the permeability of vacuum, respectively. The fields can be expressed by a vector potential $\vec{A}$ and a scalar potential Φ.

---

[1]Equations in this section are taken from [1] and constitute basic textbook knowledge.

© Springer Nature Switzerland AG 2019
T. Ostermayr, *Relativistically Intense Laser–Microplasma Interactions*,
Springer Theses, https://doi.org/10.1007/978-3-030-22208-6_2

$$\vec{\mathcal{E}} = -\frac{\partial \vec{A}}{\partial t} - \nabla \Phi \tag{2.5}$$

$$\vec{B} = \nabla \times \vec{A}. \tag{2.6}$$

With the Lorenz Gauge $\nabla \vec{A} + \partial \Phi / c^2 \partial t = 0$, Maxwells equations yield the symmetric wave equations.

$$\frac{\partial^2}{c^2 \partial t^2} \vec{A} - \nabla^2 \vec{A} = \mu_0 \vec{j} \tag{2.7}$$

$$\frac{\partial^2}{c^2 \partial t^2} \Phi - \nabla^2 \Phi = \frac{\rho}{\epsilon_0} \tag{2.8}$$

where $c = (\epsilon_0 \mu_0)^{-\frac{1}{2}}$ is the speed of light in vacuum. The simplest plane wave solution in vacuum is a sine propagating along the z-direction.

$$\vec{A} = \vec{A}_0 \cos(\omega_0 t - k_0 z + \phi_0), \tag{2.9}$$

where ω_0 is the angular frequency, $\vec{k}_0$ is the wave vector, $k_0 = \omega_0 / c$ and ϕ_0 is the initial phase. The fields $\vec{\mathcal{E}}$ and $\vec{B}$ are then given given by

$$\vec{\mathcal{E}} = \vec{\mathcal{E}}_0 \sin(\omega_0 t - k_0 z + \phi_0), \tag{2.10}$$

$$\vec{B} = \vec{B}_0 \sin(\omega_0 t - k_0 z + \phi_0). \tag{2.11}$$

Note that we only consider linear polarization in the following, since all lasers used in this work were linearly polarized. The relationship between the amplitudes is given as

$$\mathcal{E}_0 = c B_0 = \omega_0 A_0, \tag{2.12}$$

where $\mathcal{E}_0 = |\vec{\mathcal{E}}_0|$, $B_0 = |\vec{B}_0|$ and $A_0 = |\vec{A}_0|$ respectively. The intensity of the electromagnetic wave is the time-averaged magnitude of the Poynting vector $\vec{S} = (\vec{\mathcal{E}} \times \vec{B})/\mu_0$.

$$I = \langle S \rangle = \frac{\epsilon_0 c \mathcal{E}_0^2}{2}. \tag{2.13}$$

2.2   Plane Wave Interaction with a Free Electron

Before considering more complex situations, we discuss the motion of a single unbound electron in a plane electromagnetic wave, i.e. with no static fields present, without boundaries present and with an infinite transverse interaction region. The equation of motion of an electron in the electromagnetic field Eq. 2.10 with propagation along z-direction and electric field polarized in x-direction is described by the Lorentz force.

$$\vec{F} = \frac{d\vec{p}}{dt} = -e\vec{\mathcal{E}} - e\vec{v} \times \vec{B}, \tag{2.14}$$

where $\vec{p} = \gamma m_e \vec{v}$ is the electron momentum, with the relativistic Lorentz factor $\gamma = 1/\sqrt{1 - v^2/c^2} = \sqrt{1 + p^2/m_e^2 c^2}$. The photon emission by the accelerated charge is neglected in this context. A convenient notation that will be used in the following is the normalized vector potential.

$$\vec{a}(t) = \frac{e\vec{A}(t)}{m_e c}. \tag{2.15}$$

The momentum relations are from that given as

$$\frac{p_x(t)}{m_e c} = a(t), \tag{2.16}$$

$$\frac{p_z(t)}{m_e c} = \frac{a^2(t)}{2}. \tag{2.17}$$

From this relation it follows for the angle of the motion with respect to the laser propagation.

$$\tan^2 \theta = \frac{p_x^2}{p_z^2} = \frac{2}{\gamma - 1}, \tag{2.18}$$

which has been demonstrated experimentally [2]. For an amplitude of the normalized vector potential $a_0 = e\mathcal{E}_0/m_e \omega_0 c$ larger than one, the transverse oscillation is relativistic, according to Eq. 2.16. For $a_0 \gg 1$, the longitudinal motion, which is nonlinear in a_0, will start to dominate. Since a_0 has such important implications, it makes sense to introduce its relation to the intensity.

$$I\lambda^2 = a_0^2 \cdot 1.37 \cdot 10^{18} \, \text{Wcm}^{-2}\mu\text{m}^2, \tag{2.19}$$

showing that extreme intensities are required, in order to reach the first-order relativistic regime. The electron trajectories can be integrated from the momentum relations as

$$x(t) = \frac{ca_0}{\omega_0} \sin(\omega_0 t - k_0 z), \tag{2.20}$$

$$z(t) = \frac{ca_0^2}{4\omega_0}\left(\omega_0 t - k_0 z + \frac{1}{2}\sin(2\omega_0 t - 2k_0 z)\right). \tag{2.21}$$

This motion as shown in Fig. 2.1a is characterized by an oscillation along the x-coordinate at the laser frequency ω_0. In the z-direction, the motion consists of the superposition of an oscillation at twice the laser-frequency and a linear drift motion at drift-velocity.

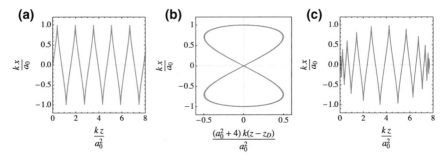

Fig. 2.1 Single electron in a plane wave. a Electron moving in a incident plane wave, observed in laboratory frame. **b** Electron moving in a incident plane wave, observed in a frame moving at the average drift velocity v_D. **c** Electron moving in a plane wave with temporal envelope described in the text, with $\tau_l = 13$ fs

$$v_D = \left\langle \frac{dz}{dt} \right\rangle = \frac{a_0^2 c}{a_0^2 + 4}. \tag{2.22}$$

Viewed from the frame co-moving with this drift motion, the trajectory is described by the well-known figure-8, shown in Fig. 2.1b. Now clearly, a laser with an infinite continuous wave is quite unrealistic. If we consider a finite pulse instead, e.g. one with a temporally Gaussian field-strength-envelope of duration τ_l and peak amplitude of $\mathcal{E}_{0,p}$, such that $\mathcal{E}_0(t) = \mathcal{E}_{0,p} \exp(-t^2/\tau_l^2)$, there is one important observation: once the laser-field is gone, the electron will find itself in another position, but at rest as shown in Fig. 2.1c. An electron would gain no energy in total, according to the *Lawson-Woodward* theorem [3, 4]. However, if just one of the conditions introduced at the beginning of this section can be overcome, energy can be gained from the laser; most high-intensity lasers are tightly focused and thus break at least the condition of infinite transverse interaction region.

2.3 Ponderomotive Potential

For an intense focused laser, the dipole approximation can become invalid, if the transverse excursion of the electron in the laser-field is of the order of the transverse intensity modulation (that is the laser-wavelength for a tightly focused laser-beam). Qualitatively one can think of an electron, sitting in the center of a laser spot with a Gaussian intensity distribution. Such electron would be accelerated away from the center, due to the electric field of the laser in the first half-cycle. When the electric field changes its sign, the electron has already traveled to a place, where the field driving it back to the center is smaller than the original field driving it out, simply due to the Gaussian geometry of the pulse. After one complete laser cycle, this results in a net offset of the electron from its original position. The process can be described quantitatively separating the fast oscillation from the comparably slow net-motion.[2] As this approach is quite universal in physics, e.g. reappearing later in the description of Paul traps, the brief derivation will be given here. In the first order of such description we neglect the effects of magnetic fields altogether and consider just a temporally cosine electric field with a radial amplitude profile.

$$\vec{\mathcal{E}}(t, \vec{x}) = \vec{\mathcal{E}}_0(\vec{x}) \cos \omega_0 t, \tag{2.23}$$

where $\vec{x}$ is the radial electron position and $\vec{\mathcal{E}}_0(\vec{x})$ is the local electric field amplitude. The equation of motion of the electron is again given by the Lorentz force Eq. 2.14. The first order acceleration of the electron is then given by

$$\frac{d\vec{v}_1}{dt} = -\frac{e}{m_e}\vec{\mathcal{E}}_0(\vec{x}_0)\cos \omega_0 t, \tag{2.24}$$

where $\vec{x}_0$ is the initial position. Integration yields the first order velocity $\vec{v}_1$ and particle position $\vec{x}_1$ as

$$\vec{v}_1 = -\frac{e}{m_e\omega_0}\vec{\mathcal{E}}_0(\vec{x}_0) \sin \omega_0 t, \tag{2.25}$$

$$\vec{x}_1 = \frac{e}{m_e\omega_0^2}\vec{\mathcal{E}}_0(\vec{x}_0) \cos \omega_0 t. \tag{2.26}$$

In the second order the effects of the magnetic field are evaluated as $\vec{v}_1 \times \vec{B}_1$. With the Maxwell equation 2.3 the first order magnetic field $\vec{B}_1$ is given as

$$\vec{B}_1 = -\frac{1}{\omega_0}\nabla \times \vec{\mathcal{E}}_0|_{\vec{x}=\vec{x}_0} \sin \omega_0 t. \tag{2.27}$$

The expansion of $\vec{\mathcal{E}}(\vec{x})$ about $\vec{x}_0$ reads.

$$\vec{\mathcal{E}}(\vec{x}) = \vec{\mathcal{E}}(\vec{x}_0) + (\vec{x}_1\nabla)\vec{\mathcal{E}}|_{\vec{x}=\vec{x}_1}. \tag{2.28}$$

Implementing above equations, the second oder equation of motion reads.

$$\frac{d\vec{v}_2}{dt} = -\frac{e}{m_e}\vec{\mathcal{E}}_0 \cos \omega t - \left(\frac{e^2}{m_e^2\omega_0^2}\vec{\mathcal{E}}_0 \cos^2 \omega_0 t \cdot \nabla\right)\vec{\mathcal{E}}_0 - \frac{e^2}{m_e^2\omega_0^2} \sin^2 \omega_0 t \left[\vec{\mathcal{E}}_0 \times (\nabla \times \vec{\mathcal{E}}_0)\right]. \tag{2.29}$$

Averaging over a full laser cycle, the cos-term vanishes and $\cos^2 \omega_0 t = \sin^2 \omega_0 t = 1/2$, simplifying the equation of motion to

---

[2]This derivation of the ponderomotive potential closely follows Reference [5].

$$\left\langle \frac{d\vec{v}_2}{dt} \right\rangle = -\frac{1}{2}\frac{e^2}{m_e^2\omega_0^2}\left[(\vec{\mathcal{E}}_0\nabla)\vec{\mathcal{E}}_0 + \vec{\mathcal{E}}_0 \times (\nabla \times \vec{\mathcal{E}}_0) \right], \tag{2.30}$$

$$\left\langle \frac{d\vec{v}_2}{dt} \right\rangle = -\frac{1}{4}\frac{e^2}{m_e^2\omega_0^2}\nabla\vec{\mathcal{E}}_0^2. \tag{2.31}$$

The net acceleration points along the gradient of the electric field towards regions of lower intensity. It can be identified with the net force, in this context referred to as the ponderomotive force $f_p = m_e\langle d\vec{v}_2/dt\rangle$. The corresponding ponderomotive potential follows as

$$\phi_p = \frac{1}{4}\frac{e^2}{m_e\omega_0^2}\vec{\mathcal{E}}_0^2 = \frac{m_ec^2}{4}a_0^2. \tag{2.32}$$

In a fully relativistic description, an additional factor $1/\overline{\gamma}$ is introduced to f_p [2, 6], where $\overline{\gamma}$ is the cycle averaged relativistic Lorentz factor. The ponderomotive potential in fact is equivalent to the cycle averaged quiver energy of the charged particle in the laser field.

As a concluding remark it shall be mentioned, that several aspects concerning the (relativistic) ponderomotive force and electron acceleration by intense focused laser-pulses, including polarization dependence in various situations, are not fully settled and have been discussed for decades (e.g. [7], and comments). One alternative possibility for the electron to gain net energy from the laser pulse, even if this is not focused, is its 'birth' during the pulse (e.g. via ionization close to the peak-intensity). The maximum energy for an electron that is initially at rest, and born into the pulse at a zero-field-crossing close to the pulse maximum, is given by Ref. ([8], pp. 18–19) as $E_m = m_ec^2(\gamma_m - 1)$ with $\gamma_m = \gamma_{t=0} - \gamma_{t=\infty} = 1 + a_0^2/2$. More generally speaking, the Lawson-Woodward theorem can break for any kind of broken symmetry in time or space domain.

2.4 Ionization

The discussion above was given for the single particle motion. Instead, this thesis is concerned with the interaction of high-power lasers with microplasmas consisting of many charged particles. Starting from the next section, discussions will disregard the process of plasma-formation, i.e. the ionization (which is an interesting subject matter in its own right, for microplasmas in particular [9]). This section will justify the further treatment of test-systems as fully ionized plasmas in the light of utilized laser peak-intensities. Since the ionization of material at lower intensities plays critical roles not just in the laser-target interaction, but already in the production and delivery of high-power laser pulses, requiring their increasing beam-diameter with beam-power, and providing the working principle of plasma mirrors, to name just two examples that will reappear later in this thesis, the very basic processes leading to ionization will be described briefly.

As a first estimate, consider the Coulomb field acting on the electron of a hydrogen atom

$$\mathcal{E}_C = \frac{e}{4\pi\epsilon_0 a_B^2} \approx 5.1 \text{ GV/m}, \tag{2.33}$$

where $a_B = 4\pi\epsilon_0\hbar/m_e e^2 = 0.53 \cdot 10^{-10}$ m, represents the Bohr radius. The corresponding laser intensity is given by Eq. 2.13 and amounts to $I_a \approx 3.51 \cdot 10^{16}$ Wcm^{-2}, often referred to as the *atomic intensity*. At laser peak intensities that were used in presented experiments, exceeding $I = 10^{20}$ Wcm^{-2}, any material in focus is ionized well before the peak of the laser-pulse arrives. In fact, much effort is required to suppress premature dynamics of the target at lower intensities (cf. Sect. 3.2).

However, already much below the atomic intensity, there exists a number of mechanisms that can ionize matter. The first, and most obvious one, is the *multiphoton ionization* [10]. Even if the single photon energy is insufficient to ionize the atom, two or more photons incident within a short time-frame can still suffice. This mechanism strongly depends on laser intensity and typically starts at intensities beyond 10^{10} Wcm^{-2}. If the electron is excited by more photons than required for ionization (so called *above threshold ionization*), it will carry the excess energy as kinetic energy, visible in the electron energy distribution in distinct peaks separated by the photon energy $\hbar\omega$ [11].

At even higher intensities, the laser starts to significantly alter the potential of the atom, giving rise to the *tunneling ionization* [12]. This occurs, when an electron can quantum-mechanically tunnel through the altered Coulomb barrier. The Keldysh parameter gives insight to the dominance of multiphoton or tunneling ionization by comparing the ionization potential V_i (13.6 eV for hydrogen) to the ponderomotive potential of the laser

$$\gamma_K = \sqrt{\frac{V_i}{2\phi_p}}. \tag{2.34}$$

For hydrogen and a wavelength of $\lambda_0 = 1$ μm the break even $\gamma_K = 1$ is found at $I \approx 7.3 \cdot 10^{13}$ Wcm^{-2}. For $\gamma_K < 1$, tunneling ionization will dominate.

At further increased intensities, the potential barrier can be suppressed even below the ionization potential of the electronic state of the atom. Then, the electron can easily escape the nucleus. This process is known as *barrier suppression ionization* [13]. It leads to a significant increase of ionization rates as compared to other mechanisms.

Note that, since the ponderomotive potential of the laser is of critical relevance in target-ionization, the ionization-degree of high-Z material is a potential direct measure for most extreme laser-intensities that are difficult to validate by other means.

If a sufficient number of electrons have been freed via primary processes, the collisional ionization can gain significance, where freed electrons ionize further atoms. The description of these processes and the prediction of average charge states in the plasma is complex, especially for systems that are not in local thermal equilibrium such as short-pulse laser-microplasmas. However, since the laser peak intensities used in this work are many orders higher than that required for any ionization mechanism (at least for carbon ions and protons), it is assumed from here on that the laser interacts with fully ionized plasmas.

2.5 Plasma

About 99% of the visible matter in the universe are in the plasma state.[3] This means, that we live in one of the rare places cold enough to allow for other states than a plasma to exist. Per definition [14], plasma is a quasineutral, partially ionized medium, exhibiting collective behavior. According to this definition, plasma consists at least in parts of charged particles; Their displacement can cause local areas with negative or positive net-charge and thereby generate long-range electric fields. Additionally, currents can cause magnetic fields in the plasma. Both kinds of fields affect charged particles over long distances, giving rise to the collective behavior of plasma, meaning that particles are not only influenced by local conditions, but also by the plasma state in more distant regions. This is particularly true for very hot collisionless plasmas, as they are discussed within this thesis, where local collisions have only negligible effects as compared to the dominating long-range forces.

Quasi-neutrality in the plasma is established by Debye shielding; particles in the plasma organize in a way that shields inner regions of the plasma from external fields. A plasma that is initially neutral contains equal numbers of positive and negative charge and is thus neutral at long length-scales. As indicated by the term quasi, neutrality can break down at smaller length-scales. A test-charge in the plasma will attract particles of opposite charge and repel those of same charge. This effect would shield the field perfectly in absence of thermal motion. For finite temperatures, some particles at the periphery of this charge-cloud can escape. Therefore, the shielding is confined to the radius at which the potential energy matches the thermal energy and fields of order $k_B T_e/e$ can leak into the quasi-neutral plasma region.

For an approximate calculation of the shielding radius, the situation can be considered in a single dimension following Ref. [14]. A charged grid spanning the plane at $x = 0$ is held at the constant potential ϕ_0 and surrounded by the otherwise quasi-neutral plasma. This situation is sketched in Fig. 2.2. Since ions have much larger inertia then electrons, they are considered as an immobile background of constant density. The Poisson equation reads

$$\epsilon_0 \frac{\mathrm{d}^2 \phi}{\mathrm{d}x^2} = -e(n_i - n_e), \tag{2.35}$$

where ϕ is the electric potential, $n_{i,e}$ are the ion and electron number densities and the atomic charge Z was set to unity for simplicity. Using the condition $n_e(\phi \to 0) = n_\infty = n_i$, where n_∞ is the density in large distance, and integrating over the electron velocity distribution, the electron number density distribution can be written as

---

[3]This section is based on textbook knowledge from [14].

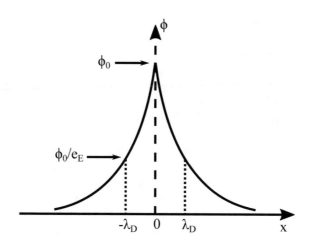

Fig. 2.2 Potential distribution near a grid in a quasi-neutral plasma. The grid at $x = 0$ is held at potential ϕ_0 and shielded by the plasma over the characteristic scale, the Debye length λ_D. Adapted from reference [14]

$$n_e = n_\infty \exp(e\phi/k_B T_e) \tag{2.36}$$

Substituting this into the Poisson equation gives

$$\epsilon_0 \frac{d^2\phi}{dx^2} = en_\infty \left(\exp(e\phi/k_B T_e) - 1\right). \tag{2.37}$$

To derive the radius of the non-neutral region, we consider regions with $|e\phi/k_B T| \ll 1$ (i.e. close to the edge of the shielding cloud), which lets us simplify the right hand side of the equation using a first order Taylor expansion. The expansion is invalid close to $x = 0$, i.e. directly at the grid, but fairly accurate towards the edge, the region of interest. It reads

$$\epsilon_0 \frac{d^2\phi}{dx^2} = \frac{n_\infty e^2}{k_B T_e}\phi. \tag{2.38}$$

This differential equation can be solved for $\phi(x)$ easily.

$$\phi(x) = \phi_0 \exp(-|x|/\lambda_D), \tag{2.39}$$

where $\lambda_D = (\epsilon_0 k_B T_e/n_\infty e^2)^{1/2}$ is the Debye length which determines the range, at which the external potential is shielded by the plasma to $1/e_E$ of its value at $x = 0$, where e_E is Euler's number. Increasing electron temperature increases λ_D, while increasing electron density results in smaller Debye length, as the shielding becomes more effective with higher electron number density. Using Debye length as the relevant length scale, the concept of quasi-neutrality can be qualified as follows: if a system is big compared to λ_D, external or intrinsic fields will be shielded over a comparably small range, and the system as a whole appears to be neutral when observed (or averaged) over lengths scales comparable to the system size. On the other hand, if the system is comparable or smaller than the Debye length, charge separation can occur over length scales comparable to the system size, and needs to be taken into account for considerations of the dynamic evolution of the whole plasma system.

Already the title of this thesis indicates the use of micro-plasmas, meaning more precisely that their size compares with the Debye length and quasi-neutrality can thus be broken. In this sense, the term 'plasma' as defined in [14] does not strictly apply to most systems described herein and shall be thought of as 'micro-plasmas' wherever suitable.

After defining these basic properties of what makes a plasma, its reaction to external force can be investigated. Any displacement of electrons from the quasi-immobile ion background will cause an electric field directed to restore neutrality. As the electrons are pulled back towards their original position, they will overshoot and start oscillating around their original position, while ions are considered to stay at rest due to their larger mass. The frequency of this oscillation is referred to as the (electron) plasma frequency, which dictates much of the plasma response to electromagnetic fields. An intuitive way to understand the plasma frequency can be found considering the Debye length beyond which the collective motion occurs, and the thermal velocity of electrons $v_{th} = (k_B T_e / \overline{\gamma} m_e)^{\frac{1}{2}}$. The response-time-scale τ_p for collective plasma motion is then given as

$$\tau_p = \lambda_D / v_{th}, \tag{2.40}$$

which is exactly the inverse of the (collision-less) plasma frequency of electrons.

$$\omega_p(n_e) = \sqrt{\frac{e^2 n_e}{\overline{\gamma} m_e \varepsilon_0}}. \tag{2.41}$$

which depends critically on the electron density. As electrons do not move freely in a plasma with ions providing a restoring force, the cycle averaged Lorentz factor is usually taken as $\overline{\gamma} = \sqrt{1 + a_0^2/2}$ [15] in the cold plasma approximation. More elaborate descriptions of the plasma dynamics can be given in the framework of the kinetic Vlasov-Maxwell equations. In special cases, derivations known as Vlasov-Poisson equations and fluid equations can yield simplified descriptions of such problems. Another option is the particle-in-cell method to simulate dynamics in the plasma. More details on the modeling of complex situations in plasmas can be found in [14] and other classic textbooks. Their relation to ion acceleration in microplasmas will be summarized in Chap. 5.

2.6 Propagation of Electromagnetic Waves in Plasma

If a relativistic laser does not interact with a single electron as in Sect. 2.2–2.3, but with a plasma target consisting of many charged particles of different kinds, the situation is complex. After ionization the further propagation of the laser in the plasma depends on the specific characteristics of the target/plasma. Derivations are given in many textbooks (e.g. [15, 16]). The collisionless dispersion relation for a transverse plane wave in the plasma reads.

$$k \approx \pm \frac{w_0^2}{c} \sqrt{1 - \frac{w_p^2}{w_0^2}}. \tag{2.42}$$

The relevant parameter here is again the plasma frequency. From Eq. 2.42 it is obvious that lasers can only propagate in plasmas with $w_0 > w_p$. Such plasmas are therefore called **underdense**. When a laser propagates through an underdense plasma, the laser pushes the electrons due to its ponderomotive force. The slow reaction of the plasma ions leads to the generation of plasma density modulations moving forward with the group velocity of the laser. In this laser wakefield, electrons can be trapped and accelerated to very high energies over very short distances which has been demonstrated by numerous laboratories around the world (e.g. [17]).

For $w_0 < w_p$ the plasma will be opaque to the laser light. Such plasmas are called **overdense**. Typically the front-surface of such a target is ionized and a pre-plasma is generated with increasing density towards the bulk target. The density where $w_0 = w_p$ is referred to as the **critical density** n_c, and follows from Eq. 2.41 as

$$n_c(w_0) = \frac{m_e \varepsilon_0 \overline{\gamma} w_0^2}{e^2}. \tag{2.43}$$

At the critical surface, where $n_e = n_c$ holds true, the laser is in parts reflected and in parts absorbed. The evanescent wave will reach into the overdense region of the plasma. For a step-like boundary we can define the characteristic scale length referred to as (relativistic) skin depth $l_s = c/\sqrt{w_p^2 - w_0^2} \approx c/w_p$, where the intensity drops to $1/e_E$ times its vacuum value. Note, that in these relations, the relativistic mass enhancement for $\gamma > 1$ is reflected in a reduced plasma frequency and an increased skin depth.

2.7 Intense Laser Absorption in Plasma

As mentioned earlier, current lasers can not directly accelerate ions to significant energies due to their large inertia. The amplitude of the normalized vector potential for ions $a_0^i = a_0 m_e/m_i$ is by at least a factor of 1836 (for protons) smaller for a given laser intensity than that for electrons. The required laser intensity for relativistic oscillatory proton motion ($a_0^i = 1$) amounts to $I^i \lambda^2 = 4.6 \cdot 10^{24}$ Wcm$^{-2}\mu$m^2. Instead, with current lasers in the range $I\lambda^2 = 10^{18} - 10^{22}$ Wcm$^{-2}\mu$m^2, the laser energy is intermediately transferred to the electrons before generating energetic ions and photons. Laser interaction with electrons is thus of particular interest for sources of secondary radiation from laser-plasmas and will eventually determine their properties and efficiency.

For moderate intensities below $I\lambda^2 \approx 10^{15}$ Wcm^{-2}μm^2, collisions between heated electrons and ions are the major cause for energy transfer from the laser to the plasma. Collisional absorption mechanisms play an important role during the early interaction of the laser pulse (e.g. via prepulses) with the target and result in preplasma expansion. Beyond this intensity, electrons will have low collision cross-sections due to their large kinetic energies. Processes leading to the absorption of such laser-pulses and consequent heating of target electrons are gathered under the notion collisionless absorption. Electron temperatures reached by collisionless absorption are typically much hotter than reached via collisional absorption mechanisms. Here we will focus on the most relevant collissionless mechanisms: **resonance absorption**, **vacuum/Brunel heating** and $\vec{\mathbf{j}} \times \vec{\mathbf{B}}$ **heating**.

Resonance Absorption

A laser pulse will generally be reflected back from the critical surface, i.e. the position in the plasma where electron density and critical density are matched $n_e = n_c$. With the pulse incident at an angle θ between its wave vector k and the plasma density gradient ∇n_e, it can reflect already at lower electron densities [5].

$$n_e = n_c \cos^2 \theta. \tag{2.44}$$

Partially the electric field can tunnel further into the plasma and interact with the critical density region. If a component of the electric field $\mathcal{E}$ of the laser pulse points in the direction of the density gradient, it drives electron oscillations and causes density fluctuations which can be resonantly enhanced. With the resulting electron plasma wave, laser energy can be efficiently coupled into the plasma. The mechanism becomes important for intensities $> 10^{15}$ Wcm^{-2}μm^2 [18].

Vacuum Heating

If the excursion-amplitude $x_o = e\mathcal{E}_0/m_e\omega_0^2$ (approximated via Eq. 2.26) of electrons oscillating along the gradient, exceeds the plasma density scale length L, the resonance breaks down. Instead, laser energy can be absorbed by the plasma via the not-so-resonant, resonant absorption mechanism, also known as vacuum- or Brunel-heating [19]. The laser field acts on electrons near the sharp plasma-vacuum interface; with an electric field component that points normal to the surface (similar to the resonance absorption), electrons can be pulled away from the surface into vacuum. In the next optical half-cycle they will be pushed back into the plasma as the laser field reverses its direction. Beyond the critical density, the laser field itself can only penetrate the target evanescently up to the skin depth l_s. However, electrons can penetrate further. Thus, once electrons cross the critical density, they are shielded from the laser field and energy is transmitted from the laser to the plasma.

j × B Heating

This mechanism arises from the oscillating $\vec{v} \times \vec{B}$ term in the Lorentz force, dominating the electron motion for relativistic laser intensities ($a_0 \gg 1$). It causes a longitudinal oscillation at twice the laser frequency $2\omega_0$ in case of a linearly polarized laser (Eq. 2.21). With a steep plasma gradient this mechanism works analogously to vacuum heating: it pushes electrons into the plasma, beyond the evanescent wave and results in plasma heating [20]. In contrast to vacuum heating, which works only by virtue of the electric field component normal to the target-surface, the laser-energy in case of the $\vec{j} \times \vec{B}$-mechanism is converted to longitudinal motion by help of the magnetic field. Thus, it performs ideal at target-normal incidence of the laser and gains overall significance once the laser reaches relativistic intensities.

Since electron heating is a stochastic process involving many particles, a complete analytical treatment is difficult. Instead, numerous scalings have been proposed and tested relying on more fundamental analytical insights and simulations [21, 22]. The most frequently used scaling [21] for flat targets [23] is the so called ponderomotive scaling. It predicts an electron temperature of

$$k_B T_e \approx m_e c^2 \left(\sqrt{1 + \frac{a_0^2}{2}} - 1 \right). \tag{2.45}$$

It relates the electron temperature to the cycle averaged energy in the quiver motion of the electron (i.e. the ponderomotive potential) via $\bar{\gamma}$. In investigations of laser-ion acceleration, such scaling facilitates estimates to relate measurements with analytical and numerical models. For instance, by use of the ponderomotive scaling we can estimate the number of hot electrons [23].

$$N_{eh} \approx \frac{\eta E_L}{k_B T_e}, \tag{2.46}$$

that are generated in the interaction. Here E_L is the laser energy, and η is the efficiency of laser-energy absorption; for laser-parameters used in our work and for bulk targets, a frequently used approximation is $\eta \approx 0.5$ [24].

The ponderomotive scaling and its details are subject to active discussions and investigations. It was criticized for its tendency to overestimate temperatures for ultrahigh intensities and flat targets ([22], and comments), while its validity for spherical mass-limited targets with size comparable to the laser focus holds in simulations [25] in several cases. For these targets and laser-pulses of sufficient duration and focal spot size, the temperature can exceed the ponderomotive scaling due to recirculation of electrons into the laser. Finally it is remarkable, that the relation of the ponderomotive scaling to the single electron motion via the cycle averaged Lorentz-factor—taken from the cold plasma approximation—may allow its extension beyond bulk plasma targets by using an adjusted single electron picture, e.g. the ponderomotive potential Eq. 2.32 for free electrons that do not remain bound to a target.

Despite not being accurate to ultimate precision, the ponderomotive scaling predicts the correct ballpark in terms of particle energy (and its scaling with laser intensity) for many scenarios including ones discussed here.

References

1. Jackson JD (1999) Classical electrodynamics, 3rd edn. Wiley, New York
2. Moore CI, Knauer JP, Meyerhofer DD (1995) Observation of the transition from Thomson to Compton scattering in multiphoton interactions with low-energy electrons. Phys Rev Lett 74:2439–2442
3. Lawson JD (1979) Lasers and accelerators. IEEE Trans Nucl Sci 26(3):4217–4219
4. Woodward PM (1946) A method of calculating the field over a plane aperture required to produce a given polar diagram. J Inst Electr Eng-Part IIIA Radiolocation 93(10):1554–1558
5. Kruer WL (1988) The physics of laser plasma interactions. Addison-Wesley
6. Quesnel B, Mora P (1998) Theory and simulation of the interaction of ultraintense laser pulses with electrons in vacuum. Phys Rev E 58(3)
7. Malka G, Lefebvre E, Miquel JL (1997) Experimental observation of electrons accelerated in vacuum to relativistic energies by a high-intensity laser. Phys Rev Lett 78(17)
8. Popov K (2009) Laser based acceleration of charged particles. PhD thesis, University of Alberta
9. Liseykina T, Bauer D (2013) Plasma-formation dynamics in intense laser- droplet interaction. English. Phys Rev Lett 110(14):145003
10. Mainfray G, Manus G (1991) Multiphoton ionization of atoms. Rep Prog Phys 54(10):1333
11. Eberly JH, Javanainen J, Rzażewski K (1991) Above-threshold ionization. Phys Rep 204(5):331–383
12. Chin SL (2004) From multiphoton to tunnel ionization. In: Lin SH, Villaeys AA, Fujimura Y (eds) Advances in multiphoton processes and spectroscopy, chap 3, vol 16. World Scientific, Singapore
13. Delone NB, Krainov VP (1998) Tunneling and barrier-suppression ionization of atoms and ions in a laser radiation field. Phys-Uspekhi 41(5):469
14. Chen FF (1984) Introduction to plasma physics and controlled fusion, vol 1. Springer Science+Business Media
15. Meyer-ter-Vehn J, Pukhov A, Sheng ZM (2001) Atoms, solids, and plasmas in super-intense laser fields. In: Mourou GA et al (ed) 1st ed, pp 167–192. Springer US
16. Gibbon P (2005) Short pulse laser interactions with matter. Imperial College Press London
17. Leemans WP et al (2006) GeV electron beams from a centimetre-scale accelerator. Nat Phys 2(10):696–699
18. Shalom E (2002) The interaction of high-power lasers with plasmas. Series in plasma physics. Institute of Physics
19. Brunel F (1987) Not-so-resonant, resonant absorption. Phys Rev Lett 59:52–55
20. Kruer WL, Estabrook K (1985) JB heating by very intense laser light. Phys Fluids 28(1):430–432
21. Wilks SC, et al (1992) Absorption of ultra-intense laser pulses. Phys Rev Lett 69:1383–1386
22. Kluge T, et al (2011) Electron temperature scaling in laser interaction with solids. Phys Rev Lett 107:205003
23. Fuchs J et al (2005) Laser-driven proton scaling laws and new paths towards energy increase. Nat Phys 2(1):48–54
24. Ping Y, et al (2008) Absorption of short laser pulses on solid targets in the ultrarelativistic regime. Phys Rev Lett 100(8):085004
25. Kluge T et al (2010) Enhanced laser ion acceleration from mass-limited foils. English. Phys Plasmas 17(12):123103

Part II
Experimental Methods

Chapter 3
High-Power Lasers

3.1 Generation of High-Power Laser Pulses

The theoretical foundations for lasers were one of Albert Einstein's many ground-breaking contributions to physics, realizing that instead of just absorbing or spontaneously emitting a photon, atoms could be stimulated to emit photons [1]. The word laser is an acronym for light amplification (by) stimulated emission of radiation.

The laser process requires at least three energy levels in a gain medium: the first (ground) state and two excited states with higher energies. In the pumping process, an atom is first elevated to the highest, instable state by transferring energy to it. This atom quickly relaxes to the second, meta-stable state. If the pump rate from the ground to the third state is larger than the decay from the second to the ground state, population inversion can be established, i.e. more atoms are in the excited second state than there are in the ground state. Photons with the proper wavelength (i.e. matching the energy of the transition between second and ground state) can then stimulate the relaxation of an atom in the second state to the ground state, going along with the emission of another photon at the same wavelength, which is spatially and temporally coherent, meaning that the electric field is perfectly in phase and the direction of emission is parallel to the stimulating photon. In the laser oscillator, the gain medium typically sits in between two mirrors. In this cavity, the coherent beam of light can travel back and forth. Every time a photon crosses the gain medium, it will be amplified by the laser process. On average, photons pass through the gain medium and get amplified multiple times, before leaving the system through the output coupling (e.g. small aperture or leakage in one mirror) or being lost to diffraction and absorption processes. If the amplification in the medium per round-trip exceeds the resonator loss, the power of the circulating beam increases almost exponentially until saturation is reached (i.e. pump energy is depleted). While the gain medium does in principle amplify any photon of the suitable wavelength running through it, i.e. regardless of its direction, only those photons in a spatial mode of the resonator geometry will run through the gain medium multiple times and thereby receive the substantial amplification.

© Springer Nature Switzerland AG 2019
T. Ostermayr, *Relativistically Intense Laser–Microplasma Interactions*,
Springer Theses, https://doi.org/10.1007/978-3-030-22208-6_3

The laser pulse duration is inherently connected to the frequency bandwidth of the laser via the time-bandwidth product. This limit is $\tau \cdot \Delta\nu > 0.44$ for the example of a pulse with Gaussian temporal and spectral profile, where τ and $\Delta\nu$ denote the FWHM values. Short laser-pulses therefore require a certain minimum-bandwidth of frequencies supported by the gain medium. Many methods exist to produce short pulses instead of a continuous wave (CW) laser-beam. Most fs-range lasers employ passive Kerr-lens mode-locking (KLM) schemes in the oscillator, where only pulses of short duration are focused by the non-linear (i.e. intensity dependent) Kerr-lens through an aperture and thus supported as a mode in the oscillator, while longer pulses are suppressed by this modified cavity.

For producing ultra-high power laser-pulses, PW class laser systems rely on the chirped pulse amplification (CPA [2], Fig. 3.1) scheme and multiple subsequent amplification stages. The initially ultrashort but relatively weak (typically nJ) laser pulse is produced by a mode locked laser-oscillator. This pulse is stretched in time by variation of the optical path of different wavelengths contained in the pulse (e.g. by insertion of a prism). Subsequently it is amplified in a number of laser-amplification stages that may have a variety of layouts to reach the desired pulse energy. Two wide spread types of amplifiers are regenerative cavities [3] that serve for additional pulse-shaping, and multipass amplifiers [4] for high energy gain in the final amplification stages.

Newer concepts (including ones used here) additionally rely on optical parametric chirped pulse amplification (OPCPA) as a part of the amplification process, combining the CPA technique with the optical parametric amplification (OPA [5]). In OPA, the signal-beam (i.e. the laser-pulse) and a shorter-wavelength pump-beam (from which the energy shall be transferred to the signal beam) travel through the same crystal. In a nonlinear process, photons of the pump are converted to equal numbers of signal-photons and idler-photons. The photon energy of the idler wave is the difference between the photon energies of pump and signal beams. Since (ideally)

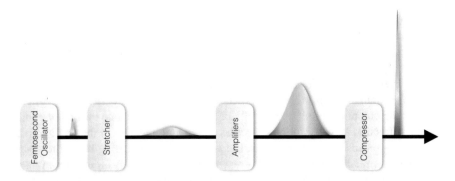

Fig. 3.1 Principle of chirped pulse amplification. The weak (nJ) femtosecond-pulse is stretched in time by 3–5 orders of magnitude and amplified over several stages. Before conducting an experiment the pulse is re-compressed in time by inverting the dispersion introduced via the stretcher

no energy is transferred to the crystal in this process, it is not heated. The OPA has another important characteristic when compared to a conventional amplifier; its spectral gain-bandwidth is not limited by laser-transitions in the gain medium, and hence allows the amplification of extremely short laser pulses even to the relativistic regime (cf. light wave synthesizer in reference [6]).

Regardless of amplification techniques, the growth in energy usually includes the growth of the laser diameter and several spatial filters in between subsequent amplification stages to maintain the desired beam profile, required to omit laser induced damage of optical components and to minimize nonlinear effects. After the final laser-amplifier, the laser pulse is re-compressed in time by inverting the optical path difference that was introduced earlier. The resulting high-energy (e.g. 100 J) and ultrashort (e.g. 100 fs) laser pulse is strong enough to immediately trigger non-linear effects when traveling in air, even when being collimated at diameters of 300 mm. For this reason, the compression, the delivery to the experiment and the experiment itself require vacuum environment. For the actual experiment, the collimated laser is delivered to an experiment chamber that contains the target system and diagnostics for the experiment. For reaching ultrahigh intensities the laser pulse is focused onto the target by a high-precision off-axis parabolic mirror allowing to focus the laser close to the physical limit of diffraction.

Example: Texas Petawatt laser The Texas Petawatt laser is an example of a hybrid system using both OPCPA and conventional CPA stages (setup sketch in Appendix A). More precisely, the large spectral gain-bandwidth of the OPCPA is used up to the 1 J level. Thus far, several tens of Joules can be considered as a rough upper limit for OPCPA stages as well as for (conventional) Ti:sapphire as a gain medium, since both need to be pumped by lasers that need to provide sufficient energy. Furthermore crystals of sufficient size for high energies are difficult to produce. Instead, similar to other early Petawatt lasers, the Texas Petawatt implements two types of Nd:glass amplifiers for the final amplification to the 150–200 J energy level. The low gain-factor in the final stages (<400) allows maintaining a bandwidth that supports compression to 150 fs. Nd:glass amplifiers can be pumped with flash-lamps and can be produced at sufficient size at comparably low cost. After compression, the laser is transported to one of two experimental chambers; the first one with a fast focus optics (f/3), which was used in experiments here, and another one using a long focus optics (f/40). Both differ in their diffraction limited spot-size (diameter of the first Airy disk) via $s = 2.44 \cdot \lambda \cdot f/\#$ and hence their highest intensity. The experimental infrastructure of the laser is displayed in Fig. 3.2.

Laser parameters It is noteworthy that the above section describes in very simplified and general terms the generation of high power laser pulses. Central parameters of laser-systems employed for experiments in the framework of this thesis are summarized in Table 3.1.

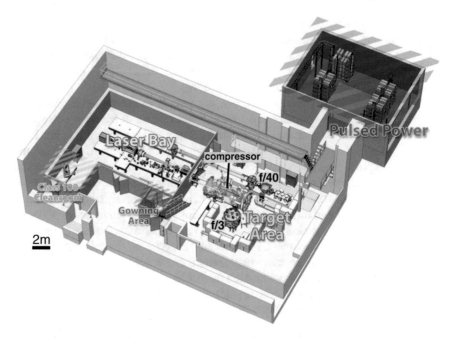

Fig. 3.2   Texas Petawatt facility. The Texas Petawatt facility and experimental infrastructure. The laser is produced and amplified in the laser bay. It is then sent through the compressor, where it is compressed to its final pulse duration before it is directed to one of the target chambers in the target area. The laser diameter at this point is 22 cm. Depending on the requirements of the experiment, target chambers with short (f/3, used here) and long (f/40) range focusing optics are available. Figure adapted with permission from [7]

Table 3.1   Laser systems. Lasers used for work presented in this thesis. The maximum repetition rate is specified for full power shots

Parameter/Laser system	TPW 2014	TPW 2016	PHELIX
Central wavelength [μm]	1.058	1.058	1.053
Pulse energy [J]	65	100	150
Pulse duration [fs]	150	140	500
Max. rep. rate [Hz]	0.0002	0.0002	0.0002

3.2   Spatio-Temporal Laser Pulse Contrast

Above descriptions of lasers do not take into account imperfections in the spatial and temporal structure of the laser. Since we use such lasers with micro-targets that are even more susceptible to such imperfections than ordinary bulk targets, they shall be considered here.

Temporal Contrast

During the generation of high power laser pulses via the CPA approach there exist several mechanisms preventing the generation of temporally perfect laser pulses. I.e. on a level that may be many orders below the peak intensity, high intensity laser pulses typically exhibit complex structures on temporal scales exceeding the pulse duration by many orders of magnitude (ns). When having the laser pulse interact with thin (nm-μm) foil targets or levitating targets in the same scale, the good laser pulse contrast is crucial. Particularly the existence of light arriving before the main pulse may cause the pre-expansion or damage of the target prior to the main pulse, via processes known from earlier days 'high-power' (GW, ns) lasers. In a worst case scenario the target will be gone at the time of peak-interaction and the acceleration of ions and other desired effects will be eliminated or strongly suppressed. Contrast must thus be controlled carefully. Typical contrast measurements use either fast diodes to measure contrast up to several ns prior to the main pulse or third order autocorrelation for measurements extending to typical $\pm$ 300 ps.

Amplified spontaneous emission, pre-pulses and coherent contrast Major sources of light arriving prior to the main pulse are the amplified spontaneous emission (ASE), pre-pulses and the so called coherent contrast. The ASE process sets in as soon as amplifier crystals are pumped to inversion. The luminescence can be amplified through the amplifier chain and generate a quasi-constant pedestal typically in the ps to ns-timescale. The ASE process is inherent to laser amplifiers and may be suppressed to a certain level by optimizing the laser system with this regards, e.g. by use of fast Pockels cells. In OPA and OPCPA, the parasitic fluorescence noise (originating from the OPA pump pulse time window) can have a similar effect on the temporal contrast as the ASE.

In addition, short pre-pulses can be introduced to the temporal structure of the pulse, generally via surfaces of optical elements (e.g. Pockels cell windows or lenses) that reflect parts of the light in the laser propagation direction. Via spectral phase modulation even post-pulses generated by such surface may turn into pre-pulses during the pulse re-compression and may therefore not be neglected. One key to a good laser contrast is thus the identification and elimination of flat surfaces in the laser chain that produce significant pre- and post-pulses.

The third source of premature light is the coherent contrast that starts to rise few ps to tens of ps before the main pulse. As of now, the source of this is still being discussed. In some stretcher/compressor geometries the spectral clipping may contribute. Another source is the spectral phase noise introduced by imperfect optics that hinder the compression to the Fourier limit [8–10].

Requirements on temporal laser contrast The required temporal laser contrast is mainly dictated by two parameters. The first parameter is the peak intensity of the laser pulse, say 10^{20} W/cm^2 for example. The second parameter is the damage threshold intensity of the target material at the specific laser-pulse duration, typically 10^{13} W/cm^2 for plastics. In that example a laser pulse contrast of 10^{-7} would be the minimum requirement, at least up to several tens of ps before the peak pulse.

While best results for laser ion acceleration are not necessarily obtained with the best laser contrast, the best possible laser contrast is beneficial in order to assure controlled target parameters during the peak-interaction. This simplifies insights to the laser-plasma interaction in simulations and models by pinning down laser and target parameters, which are otherwise extremely challenging to monitor during peak-interaction.

For this reason, many laser systems have incorporated special elements to enhance the laser contrast. Techniques include Pockels cells (electro-optical elements) and saturable absorbers [11] to suppress ns prepulses and ASE as well as cross-polarized wave generation (XPW [12]) and optical parametric CPA (OPCPA) with ps pump durations for an improved contrast on the ps timescale.

Plasma mirrors One of the most widespread techniques to enhance the laser contrast is the plasma mirror. This is an ultrafast laser-triggered optical shutter that improves laser contrast by two to three orders of magnitude (for a single plasma mirror) in ideal conditions. The working principle is illustrated in Fig. 3.3. The high power laser including its prepulses is focused to a defined size onto an anti-reflective (AR) coated substrate (formerly known as window). This window stays transparent during the passage of prepulses/pedestal. During that time only a tiny fraction depending on the quality of the coating (that are the typical two to three orders of magnitude) is reflected in the direction of the target. At some point of the rising edge of the laser pulse the intensity will suffice to ignite a solid density plasma. For the fundamental laser wavelength a solid density plasma reflects most of the incoming laser light, and the main pulse is thus efficiently directed towards the target.

The use of plasma mirrors for the contrast enhancement of optical laser pulses has first been demonstrated in 1991 [13]. While plasma mirrors are useful tools for optimizing the laser contrast they also bring some challenges. First the substrate needs to be replaced after every shot since the AR coating is single-shot damaged. This must be done without moving the laser axis, which is even more important when focusing the laser onto isolated targets in the size range of the laser focal spot itself. The fluence on the plasma mirror shall be around 50–100 J/cm^2 for optimum conditions which at a given laser energy dictates the substrate size [13–15]. Particularly at high energy lasers, plasma mirrors become expensive. One of the largest disadvantages is the loss of energy introduced by plasma mirrors that depends heavily on the reflection angle

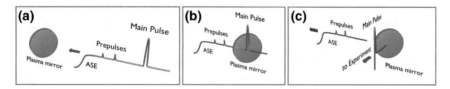

Fig. 3.3 Principle of plasma mirrors. a The re-compressed pulse including ASE and prepulses is sent towards the plasma mirror. **b** While ASE and prepulses pass the plasma mirror without deflection, the main pulse quickly ignites a plasma on the mirror surface. **c** The high density plasma reflects the main pulse towards the experiment

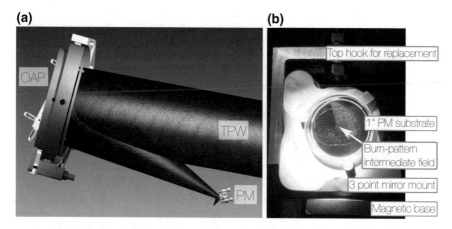

Fig. 3.4 Single inline plasma mirror at TPW. a The single inline plasma mirror as employed for the Texas Petawatt campaigns. As a scale, the initial beam diameter is about 22 cm and the plasma mirror substrate diameter is 25.4 mm. **b** Plasma mirror after a single laser-shot. The imprint of the laser-beam profile is clearly visible

and the fluence on the substrate. Typical reflectivities range from 90% downwards [13–15] and expensive laser energy is lost as a consequence, which however is often accepted in turn for the improved contrast.

A single plasma mirror implementation as depicted in Fig. 3.4 has been designed and used for both TPW experiments presented later in this thesis. In our experiments, the laser pulse was focused by the f/3 off axis parabolic mirror which is inherent to the setup. At a distance of 35 mm before the actual laser focus (at a spot size close to 1 cm^2), a double-side anti-reflex coated window of 1 in. diameter and 5 mm thickness was positioned at 45° with respect to laser propagation direction, redirecting the laser by 90°. Optionally an (initially) high-reflectivity dielectric or silver mirror could be positioned in the same position to emulate the effects of the imperfect original laser contrast. The reflectivity was measured as 80 ± 5%, which is a typical value for plasma mirrors operated in the presented parameter regime. As described earlier, using the plasma mirror we expect the laser intensity contrast to enhance by 2–3 orders until it ignites. The substrate requires replacement prior to each full-power laser shot due to substrate damage. The special three-point mirror mount in combination with a magnetic base perfectly define the mirror position and alignment inside the setup allowing for convenient substrate replacement without any measurable variation in position or pointing of the laser pulse. A manual vacuum feed-through and vacuum gate valve enable quick replacement of the substrate, even without venting the entire experimental vacuum chamber between single laser shots.

Temporal contrast at TPW Fig. 3.5 shows the normalized ns range energy contrast of the Texas Petawatt, measured before our first experiment. Even before the OPA parasitic fluorescence noise sets in at few ns before the peak, several prepulses up to $10^{-7} - 10^{-6}$ are registered. Prepulses were identified with reflections from lenses,

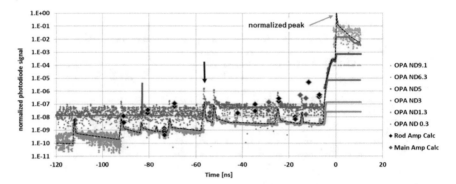

Fig. 3.5 Nanosecond energy contrast of the Texas Petawatt laser in 2014 (without plasma mirror). Photodiode traces of the compressed OPCPA pulse at multiple attenuation levels. The black dashed line provides a guide to the eye to read the measurement. Pulses at the round trip time of the amplifier are indicated by arrows. Prepulses expected from the rod- and main-amplifier design are indicated by diamonds; they appear slightly overestimated. This data is presented by courtesy of Dr. Erhard Gaul (adapted with permission from [16])

and have a different mode (mostly pencil beams [16, 17]) than the main pulse. They are expected to be less amplified, less compressed in time and less focused through spatial filters and on target than the main pulse. As a result, the intensity contrast on target is estimated to be two to three orders better than the measured energy contrast. The plasma mirror results in a further increased intensity contrast of 10^{-11}–10^{-10} up to the start of OPA parasitic fluorescence at several ns before the peak.

For a closer look at the ps region around the peak interaction, a single-shot third-order autocorrelation measurement of the laser intensity contrast in the 100 ps range is displayed in Fig. 3.6. All signal up to the rising edge of the actual main pulse (ps range) was either identified as measurement artifact, or found close to the noise level

Fig. 3.6 Picosecond intensity contrast of the Texas Petawatt laser in 2014 (without plasma mirror). The (single-shot) third order autocorrelation reveals the intensity contrast in a hundred-ps-scale range. This data is presented by courtesy of Dr. Erhard Gaul (reproduced with permission from [16])

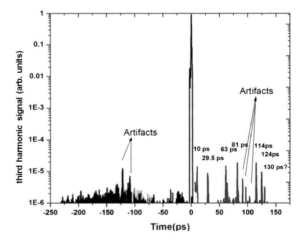

(around $2 \cdot 10^{-6}$) of the measurement [16]. The intensity contrast with the plasma mirror implementation is consequently expected to be around 10^{-8} up to the ps-level before the peak pulse, sufficient to keep a polymer target intact during that time with 10^{20} Wcm^{-2} peak intensity. The contrast discussed here is relevant to our first experiment at TPW (Sect. 5.3).

Before our second experiment at TPW (Chap. 6), a major upgrade eliminated prepulses and reduced the OPA parasitic fluorescence by replacement of all lenses with reflective optics and implementation of ps short-pulse pumped OPCPA stages (instead of ns-pumped) respectively [18]. The overall contrast was then $5 \cdot 10^{-8}$ up to the ps-level before the peak pulse, which would have—in principle—allowed to operate even without the plasma mirror, which was implemented anyway for an optimized contrast.

Spatial Contrast: Laser Focusing and Pointing Stability

In order to reach highest intensities at the target, the laser beam is focused by a reflective optic. Typically one considers ideal conditions, e.g. a perfect flat-phase, and Gaussian intensity distribution for the laser beam in the near-field, that propagates from amplifier stages, through the compressor and laser beam delivery to the perfect focusing off-axis parabolic mirror (OAP). In that case, the far-field distribution of the laser is trivially given as the Fourier transform of the perfect input distribution (in case of a Gaussian, that is a Gaussian). In reality though, the wavefront and intensity distributions will deviate from perfection due to imperfect pump profiles, gratings, mirrors, plasma mirrors and OAP. The focus quality is thus an important measurement when it comes to the interpretation of experimental results, in particular with regards to the peak intensity and corresponding scalings of relevant observables, such as particle energy distributions.

In many experiments and large parts of literature the focus quality, FWHM focus diameter and laser intensity are qualified solely by single microscope images of the focus with signal to noise ratios of 10 bit or less. During our experiments, it was recognized that such measurement is often insufficient to obtain meaningful values of laser-intensities and energy enclosed within FWHM (or within the highest intensity region, e.g. first Airy disk). This is because outside of the laser-focus, wings are produced due to optics imperfections, often carrying significant amounts of the laser energy, spread out in a large area and thus not recognized in low dynamic range (LDR) images, due to their low intensity [19].

In order to demonstrate the issue, we present a high dynamic range (HDR) focus image taken at the TPW in September 2016 with the focus diagnostics microscope. It was generated by combining four single 8bit images from single shots taken at different laser-filter settings, into a single HDR image. Intensity at each pixel is specified by partition of the laser energy (80 J behind the plasma mirror) according to the pixel-value and division through the laser-pulse duration (140 fs).

The FWHM diameter measured in the HDR image is similar to the value measured for a single non-saturated 8 bit image, about 5–6 μm. This is expectable, since the highest 50% in terms of intensity can well be recorded within the first 8 bit frame. However, the peak-intensity and FWHM intensity are reduced by a factor of 3–4 in HDR as compared to a LDR image evaluation (true peak intensity of $5.8 \cdot 10^{20}$ Wcm^{-2} instead of $1.8 \cdot 10^{21}$ Wcm^{-2}), due to the energy spread-out in the outer regions. This energy is lost in signal to noise ratio when using an LDR image and thus not taken into account properly in the intensity calculation. The discrepancy in intensity measurements with HDR and LDR images has been found not only at TPW, but with similar measurements by our group at our local 'laboratory for extreme photonics' (LEX-photonics), at the PHELIX laser (GSI Darmstadt) and at the Astra Gemini laser (Central laser facility). Recent experiments involving optical probing at LEX demonstrated that the energy in the outer regions is indeed present and not artificially created by the measurement-technique (e.g. in the microscope). The imperfect focus can thus be regarded as the spatial analogue to the temporal laser contrast, i.e. as the appearance of light in a region that would not be illuminated (or at least less illuminated) in case of a perfect laser pulse. This observation highlights the usefulness of small isolated targets for the laser-plasma interaction: they allow to record clean data from a single position (and hence intensity) in the focal plane. This is in stark contrast to the usual extended bulk foil target, which produces many simultaneous signals from a plethora of processes taking place at different intensities occurring in the focal plane (Fig. 3.7).

Another important consideration when dealing with targets of reduced dimensions is the stability of laser pointing. This means the shot to shot variation of the position of highest intensity, which can change due to vibrations of the building or due to vibrations of individual optics. As shown in Fig. 3.8 the overall pointing stability of the TPW during experiments was comparable to those of many high power lasers

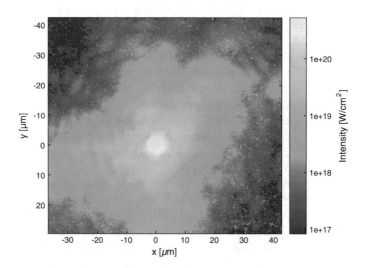

Fig. 3.7 Focus image of TPW. A high dynamic range focus image of TPW recorded in September 2016. The image was combined to an HDR image from 4 separate 8 bit images taken at different filter settings for the laser. The wings of the laser carry significant parts of the laser energy and surpass the relativistic threshold of 1.37×10^{18} $Wcm^{-2}\mu m^2$ in an area of $\sim$50 μm width

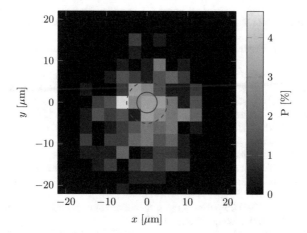

Fig. 3.8 Pointing stability of TPW. The pointing stability of the laser is evaluated with the focus diagnostic microscope, using 200 consecutive shots of the attenuated laser system. The distribution shows the probability for the peak intensity to hit in a specific region. Red-dashed and solid red circles mark 10 and 5 μm diameter. Enclosed probabilities allow considerations regarding success-rate when using according FWHM focus diameters

available to date, i.e. comparable to the focal spot size itself. Considering a target situated at the origin, the red-dashed and red circle can give some insight with regards to the probability of hitting the target center with at least half of the laser peak intensity depending on the FWHM laser focal spot diameter. For a 10 μm focal spot diameter as used in the 2014 campaign, this probability sums up to ∼20%. For a 5 μm focal spot as used in the 2016 campaign the probability is only ∼5%. It shall be noted here, that the consideration is a limiting case, since it does not take into account the spatial extent of the target and non-central laser-hits, which depending on the target-dimension can improve the total integrated probability. An important note is, that focus quality did not suffer from pointing fluctuations, i.e. aberrations due to slightly misaligned off-axis-angles caused by pointing were not observed.

References

1. Einstein A (1917) Zur Quantentheorie der Strahlung. Phys Zeitschrift 18:121–128
2. Donna Strickland and Gerard Mourou (1985) Compression of amplified chirped optical pulses. Opt Commun 56(3):219–221
3. Murray JE, Lowdermilk WH (1980) ND:YAG regenerative amplifier. J Appl Phys 51(7):3548–3556
4. Lowdermilk WH, Murray JE (1980) The multipass amplifier: theory and numerical analysis. J Appl Phys 51(5):2436–2444
5. Giordmaine JA, Miller RC (1965) Tunable coherent parametric oscillation in LiNbO$_3$ at optical frequencies. Phys Rev Lett 14:973–976

6. Rivas DE et al (2017) Next generation driver for attosecond and laser-plasma physics. Sci Rep 7(1)
7. Texas PetawattWebsite. http://texaspetawatt.ph.utexas.edu/overview.php. Last accessed 07 Nov 2017
8. Christophe Dorrer and Jake Bromage (2008) Impact of high-frequency spectral phase modulation on the temporal profile of short optical pulses. Opt Express 16(5):3058–3068 Mar
9. Hong K-H et al (2005) Generation and measurement of 108 intensity contrast ratio in a relativistic kHz chirped-pulse amplified laser. Appl Phys B 81(4):447–457
10. Hooker Chris et al (2011) Improving coherent contrast of petawatt laser pulses. Opt Express 19(3):2193–2203 Jan
11. Sung JH et al (2014) Enhancement of temporal contrast of high-power femtosecond laser pulses using two saturable absorbers in the picosecond regime. Appl Phys B 116(2):287–292
12. Minkovski N et al (2002) Polarization rotation induced by cascaded third-order processes. Opt Lett 27(22):2025–2027 Nov
13. Kapteyn HC et al (1991) Prepulse energy suppression for high-energy ultrashort pulses using self-induced plasma shuttering. Opt Lett 16(490)
14. Doumy G et al (2004) Complete characterization of a plasma mirror for the production of high-contrast ultraintense laser pulses. English. Phys Rev E 69(2):026402
15. Lévy A et al (2007) Double plasma mirror for ultrahigh temporal contrast ultraintense laser pulses. Opt Lett 32(3):310
16. Gaul E et al (2014) Pulse contrast measurements of the Texas Petawatt laser. Res Opt Sci, OSA Tech Dig JW2A 23
17. Murray JE, Van Wonterghem B, Seppala L (1995) Parasitic pencil beams caused by lens reflections in laser amplifier chains. OSTI ID: 108086, UCRL-JC–121125, CONF-9505264–17
18. Gaul E et al (2016) Improved pulse contrast on the Texas Petawatt Laser. J Phys: Conf Ser 717:012092
19. Key MH (2001) Atoms, solids, and plasmas in super-intense laser fields. In: Mourou GA et al (ed) pp 147–166. Springer US

Chapter 4
Transportable Paul Trap for Isolated Micro-targets in Vacuum

In this section, an introduction to electrodynamic quadrupole traps is given and the experimental realization is presented.[1] The theoretical part is mostly based on textbook knowledge from [6, 7].

4.1 Short Theory of Electrodynamic Traps

The Earnshaw theorem [8] states that trapping (e.g. levitation) of particles is unachievable for electrostatic fields. In contrast, time-dependent electric fields can generate ponderomotive wells and thereby enable the construction of so called Paul traps of which different kinds exist.[2]

The electric potential in a harmonically driven linear Paul trap can be written in terms of time t and two spacial dimensions x and y as

$$\phi(\vec{x}, t) = V \cos \Omega t \cdot (x^2 - y^2)/r_0^2, \tag{4.1}$$

where V is the AC voltage amplitude at the trap electrodes, Ω is the angular drive frequency and r_0 is the distance of the trap center to the electrode. Electrode surfaces should ideally follow the equipotential surfaces of the associated electric field. i.e.

---

[1]The development of this setup and related methods to their current status has been a long-term collaborative effort. It took work of five LMU master/diploma thesis (Hilz [1], Ostermayr [2], Haffa [3], Singer [4] and Gebhard [5]) of which I (co-)supervised the last three, and two dissertations (T. Ostermayr and P. Hilz), to get there.

[2]This section is reproduced with small variations, and with permission, from the original peer-reviewed article: T.M. Ostermayr et al., Review of Scientific Instruments, **89**:013302, (2018). The article is published by the American Institute of Physics and licensed under a Creative Commons Attribution 4.0 International License (https://creativecommons.org/licenses/by/4.0/).

© Springer Nature Switzerland AG 2019
T. Ostermayr, *Relativistically Intense Laser–Microplasma Interactions*,
Springer Theses, https://doi.org/10.1007/978-3-030-22208-6_4

the ideal trap would consist of four infinitely-sized hyperbolically-shaped electrodes around the trap center as depicted in Fig. 4.1a, that are infinitely extended in the third spatial dimension z, such that they cause any electric field in that direction. The axial confinement is taken care of by additional measures, e.g. additional electrodes.

The equation of motion of a particle in this potential reads as $M\mathrm{d}^2\vec{x}/\mathrm{d}t^2 = -Q\vec{\nabla}\phi(\vec{x})$, with charge Q and mass M of the particle. They take the form of Mathieu differential equations [9] for x and y

$$\frac{\mathrm{d}^2 x}{\mathrm{d}\tau^2} + 2xq\cos 2\tau = 0, \tag{4.2}$$

$$\frac{\mathrm{d}^2 y}{\mathrm{d}\tau^2} - 2yq\cos 2\tau = 0, \tag{4.3}$$

$$\frac{\mathrm{d}^2 z}{\mathrm{d}\tau^2} = 0, \tag{4.4}$$

where $\tau = \Omega t/2$ and

$$q = 4QV/Mr_0^2\Omega^2. \tag{4.5}$$

Mathieu equations are well known in terms of their solutions' behavior [6]: stable solutions remain limited and can thus be identified with a trapped particle, while instable solutions diverge and corresponding particles leave the trap after short time. The distinction can be made based on a single quantity, the q-parameter Eq. (4.5). Values of $0 < |q| < 0.908$ result in stable trajectories, larger q leads to unstable solutions and particles are not trapped.

An example of a stable solution for one set of initial conditions is depicted for one coordinate in Fig. 4.1b. The particle trajectory here resembles a superposition of two parts [10]: a fast oscillatory term, the so called micromotion x_{mic}, which is directly excited by the fast alternating electric field, and a slower motion x_{mac}, referred to as macromotion, which can be regarded as the net-motion after averaging

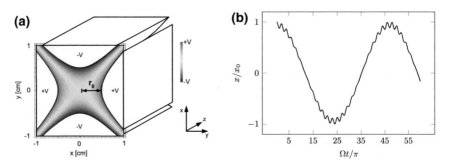

Fig. 4.1 Paul trap potential and 1D trajectory. a Snapshot of the two-dimensional quadrupole potential (contour) and one possible set of trap electrodes (white). **b** The motion of a particle in x-dimension with $q = 0.12$ and initial conditions $x(0) = x_0$, $\mathrm{d}x/\mathrm{d}t(0) = 0$, and $\mathrm{d}^2x/\mathrm{d}t^2(0) = 0$. The motion consists of two superimposed oscillations

over one trap-voltage oscillation period. The parameter q is also a good measure for the adiabaticity of the system: for parameters $|q| < 0.3$, the system can be assumed to be adiabatic [11], i.e. the energy of the macromotion is conserved and the particle is not kinetically heated by the fast oscillations of the trap's AC drive field. Then, the particle motion can be well described as a superposition of micro- and macromotion. In our system we aim to stay within this well behaved adiabatic regime of operation.

The macromotion x_{mac} is as a harmonic oscillator

$$\ddot{x}_{mac} + x_{mac}\omega_{sec}^2 = 0, \tag{4.6}$$

oscillating with the secular frequency $\omega_{sec} = q\Omega/\sqrt{8}$. The corresponding quasistatic potential reads

$$U_p(x_{mac}) = \frac{1}{4}qV\frac{x_{mac}^2}{r_0^2}. \tag{4.7}$$

This potential must be distinguished from an electrostatic potential, since it does not fulfill the Laplace equation. For this reason, it is usually referred to as a pseudopotential or effective potential in context with particle traps, since it regardless describes the time-averaged net force acting on the particle. This description provides important insight regarding the strength of the confinement in the trap [12]. Note that the pseudopotential discussed here is related in meaning, derivation and origin to the ponderomotive potential discussed earlier in context with laser-matter interactions.

In our setup, we use a trap with four cylindrical rods as AC electrodes. In contrast to infinitely-sized hyperbolically-shaped electrodes, this provides the required optical access to the trap center. On the other hand, cylindrical rod geometry introduces higher order potentials with the next higher orders being the 12-pole and the 20-pole [13]. Our arrangement deviates from the optimized cylindrical geometry [14, 15] with $r_{rod}/r_0 = 1.14511$ and vanishing 12-pole term. The resulting equations of motion in such a trap are (in principle) coupled in coordinates, and the particle motion is anharmonic, i.e. the particle's motion frequency is amplitude dependent [16]. Higher order multipoles also lead to resonances within the stable region $0 < |q| < 0.908$, causing unstable trajectories within that otherwise stable interval [17–19]. Both effects are mostly relevant to the effective trapping volume for storage of large ion clouds and their implications for precision measurements [20]. Here, we minimize the practical influence of higher order potentials by several measures: First our trap usually operates with $|q| < 0.4$, i.e. where only very high order resonances are present [19]. Second, the levitating target's position is very close to the trap center, where higher order contributions are naturally small, due to their scaling with higher orders of distance from the trap center, than the quadrupole's scaling. The difference of the quadrupole in the real trap to that of an ideal trap are taken into account simply via a modified (effective) inner radius [20] r_0^{eff} within the classical equations.

We achieve the confinement in the third spatial dimension by application of additional electrostatic potentials via end-cap electrodes. The particle motion along this coordinate resembles an oscillation that occurs at the frequency ω_z. Due to the symmetry, additional anharmonic electrostatic fields, caused by the additional electrodes, vanish in the trap center and are neglected in this discussion, where particles are confined close to the trap-center.

4.1.1 Active Trajectory Damping

Our central objective is the preparation of a trapped particle exactly in the trap center without significant residual motion, which requires damping of the trajectory of the trapped particle.[3] The amplitude of the micromotion is proportional to the amplitude of the local electric field. It thus vanishes for a particle located exactly in the trap center. Under assumption of negligible gravitation and electrostatic forces, the efficient damping of the harmonic macromotion via reduction of its amplitude is therefore sufficient for target positioning. To that end, we use the electric feedback field $\mathcal{E}(x_{mac})$ reacting to the opto-electronically measured and electronically filtered particle trajectory x with a proper phase-shift, modifying the harmonic macromotion described by Eq. (4.6) to that of a damped harmonic oscillator. For the further discussion we include arbitrary external forces F that act on the system, and the Eq. (4.6) reads

$$\ddot{x} + \omega_{sec}^2 x = \frac{Q}{M}\mathcal{E}(x) + \frac{F}{M}, \tag{4.8}$$

where we dropped the index for the macromotion for readability.

This allows the analysis of the system response to an arbitrary external force $F(t)$, which is most conveniently done in the Laplace domain. The Laplace domain, in contrast to the Fourier domain, uses damped/amplified sinusoids as eigenfunctions of the time-domain signal, and is hence well suited for the analysis of the transient system response to external forces. First we take care of the feedback function $\mathcal{E}$, which in the Laplace domain can be written by means of a transfer-function $H_f(p)$ that describes the reaction $\hat{\mathcal{E}}$ to the input value $\hat{x}$

$$\hat{\mathcal{E}}(p) = H_f(p)\hat{x}(p), \tag{4.9}$$

where the hat denotes the Laplace transform of respective functions, $p = \alpha + i\sigma$ is a complex number and $\alpha, \sigma \in \mathbb{R}$. This implicates that the feedback reacts to the measured position of a trapped particle. The system presented here uses the response of a low-pass filter. We will demonstrate the damping behavior for a first order low pass with its known transfer-function

$$H_f(p) = \frac{A}{1 + p/\omega_c}, \tag{4.10}$$

---

[3]This section is based on private communication with Ivo Cermak, CGC Instruments, Chemnitz.

where A is the dc gain of the channel and ω_c is the cut-off frequency. Most importantly the low pass adds a phase shift to the input signal which is essential for the damping mechanism to work. After transforming the entire Eq. (4.8) to the Laplace domain and implementing the low pass filter Eq. (4.10) to the electric feedback field via Eq. (4.9), we can deduce a system response function $H_s(p)$ which in a similar way describes the system's (trapped particle's) response $\hat{x}(p)$ to the external force $\hat{F}(p)$ in the Laplace domain as

$$\hat{x}(p) = H_s(p)\hat{F}(p)/M. \tag{4.11}$$

Some calculus leads to the system response function

$$H_s(p) = \frac{1}{p^2 + \epsilon p \omega_{sec}/\omega_c + (1 - \epsilon)\omega_{sec}^2}, \tag{4.12}$$

where $\epsilon = QA/M\omega_{sec}^2$ shall be much smaller than $\omega_c^2/\omega_{sec}^2$ to stay in the weak damping regime and for the equation to be valid. From that the reaction of the trapped particle to an external force $F(t)$ can be inferred by following the above steps, working in the Laplace domain, and re-transforming the result to the time-domain. Here the process is demonstrated for a short-time impulse on the system, $F(t)/M = \delta(t)$, e.g. corresponding to a short kick given to the particle at $t = 0$. The Laplace transform of the force reads $\hat{F}(p)/M = 1$ and the system reaction in the Laplace domain is therefore given by the transform function $H_s(p)$ alone, as given in Eq. (4.12). The re-transformation of the system yields the particle behavior in the time domain

$$x(t) = \frac{1}{\omega_d} \exp(-\zeta \omega_n t) \sin(\omega_d t). \tag{4.13}$$

This can be identified as a damped harmonic oscillator with the natural frequency $\omega_n = \omega_{sec}\sqrt{1 - \epsilon}$, the damping factor $\zeta = \omega_{sec}\epsilon/(2\omega_c\sqrt{1 - \epsilon})$ and the damped oscillation frequency for a weakly damped system $\omega_d \approx \omega_{sec}(1 - \epsilon/2)$. Thus, once the particle is displaced from the trap-center by a force, its general behavior is to reduce the amplitude of its natural oscillations. Effective damping usually takes few hundreds of oscillation periods, or several seconds.

4.2   Test Setup Without High Power Laser

During the development of the trap, we built a fully functional test-setup (Fig. 4.2) which did not (yet) implement the high-power laser.[4] A central part thereof is the vacuum chamber enabling operation at pressures down to 10^{-6} mbar. The typical

---

[4]This section including figures is reproduced with small variations, and with permission, from the original peer-reviewed article: T.M. Ostermayr et al., Review of Scientific Instruments, **89**:013302, (2018). The article is published by the American Institute of Physics and licensed under a Creative Commons Attribution 4.0 International License (https://creativecommons.org/licenses/by/4.0/).

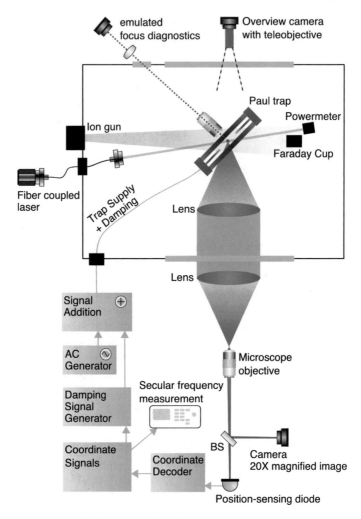

Fig. 4.2 Test setup for the Paul trap. The minimum Paul-trap test-setup comprises a vacuum chamber, the trap and connected power supplies, an ion gun to charge the target particles, a laser to illuminate target particles, and the optical measurement and feedback setup. Figure and caption reproduced with permission from [21]

laser-plasma experiment operates at pressures below 10^{-5} mbar. This is necessary in order to avoid the gaseous environment's influence on the high-intensity laser itself, and on the charge state of particle beams generated in the interaction (due to a limited mean free path). This latter point is particularly important if charged particle beams are diagnosed by electro- and magneto-static spectrometers, such as used in this work. The Paul-trap is depicted in Fig. 4.3. It consists of four cylindrical polished copper cylinders with diameters of 5 mm, arranged around the trap center at the geometric distance $r_0 = 8.1$ mm. The corresponding effective inner radius [20]

Fig. 4.3 Paul trap. The
Paul-trap, consisting of four
AC electrodes and two DC
endcaps. All electrodes are
positioned in precision
manufactured isolating
ceramics seats. Figure and
caption reproduced with
permission from [21]

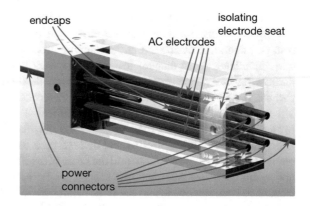

is $r_0^{\mathrm{eff}} = 8.7\,\mathrm{mm}$. This setup provides a wide optical access to the trap center. In the
current configuration, the trap can accommodate even f/1 optics.

Two round copper slabs with diameters of 3 mm serve as endcaps. They are aligned
on the $x = y = 0$ axis and mirrored at the trap center, with a distance of 10–20 mm
in between each other. All AC electrodes and endcaps are positioned with respect
to each other in precision-manufactured ceramics seats, which also isolate them
electronically from each other. All AC electrodes and both endcaps are connected to
individual voltage supplies. The setup features frequencies $\Omega/2\pi$ up to 5 kHz and
voltages V up to 3 kV for the AC electrodes and DC voltages up to 450 V for the
endcaps [22].

Above the trap center, the target reservoir is positioned. It is filled with few mil-
ligrams of particles from a commercial monodisperse sample of spherical particles
[23]. The reservoir mechanism works similar to a salt shaker: mechanical actuation
of the reservoir causes several particles to fall through a 100 μm hole into the trap.
The charging of the particles is performed with an ion beam that crosses their path.
The ion gun was built at *Tectra* [24] delivering ions with energies of up to 5 keV and
current densities of several 10 nA/mm^2 in a beam of ~20 mm (FWHM) cross section
(measured where the beam crosses the particle's path). The current can be monitored
by a Faraday cup. The ion gun uses an adjustable gas flow of air or helium, which
temporarily increases the pressure inside the chamber to 10^{-5}–10^{-4} mbar. Typically
in this initial trapping procedure, a few particles are trapped. By reducing the endcap
voltages to a few volts before the trapping, the Coulomb repulsion between trapped
particles can reliably and quickly push all but one particle out of the trap. After this
initial trapping, the ion gun is turned off, and the lowest attainable pressure is re-
established within a second. The endcap voltages are ramped up to their final value.
Overall, the initial trapping process takes around 10–30 s.

For the illumination of the trapped particles we use a 660 nm laser diode with
a max. output power of 50 mW, of which a maximum of 32 mW is coupled into a
single-mode fiber. The light is brought into the chamber with a vacuum-feedthrough.
Inside the chamber, an adjustable 'collimator lens' loosely focuses the laser from the
fiber exit to a spot size between 0.25 and 1 mm FWHM on the target. The stray light

from the target is collected by three imaging systems that serve different purposes. The overview camera uses a teleobjective and observes the trap-system and the reservoir operation during the initial trapping process. A 20X microscope, referred to as 'emulated focus diagnostics', produces micrometer-resolution videos of the particle trajectory. This serves to quantify the position and the stability of the damped particle. It resembles the microscope used in experiments for target-focus alignment. The main optical system consists of two identical 4" lenses with an F-number of 1 to transport a 1:1 image of the target to the outside of the chamber. This image is magnified by a 20X objective and divided by a beamsplitter onto: (a) a camera monitoring the particle position until the firing of the high-power laser-shot. Note that during this shot, the focus diagnostics that is used to establish the laser-target overlap, must have been removed from the laser-beam path in order to avoid its destruction. And (b) onto a position-sensing diode [25] (PSD). This central element tracks the center-of-mass motion of the particle's magnified image. Two electric signals from the PSD provide information about the projection of the particle motion onto the PSD (i.e. vertical and horizontal direction). The effective bandwidth of the coordinate detection is around 10 kHz, enabling real-time measurements of the particle motion (typically in the sub-kHz range). From the time-domain signal, the frequency of the particle motion is derived via Fourier analysis; the secular frequency is represented by the highest peak in the frequency spectrum. The same coordinate signals are used for the active electric damping of the particle motion. The particular coordinate signal was phase-shifted and applied as an additional voltage to the corresponding trap electrodes as described in the section above. The result was the effective damping of the particle motion without any gas agent or other damping method typically used to damp particle oscillations.

Note that despite using only a 2D projection on the PSD, the signal contains the information about the particle's motion along the three major axes of the trap. The vertical PSD signal contains the information on the motion in the vertical dimension of the trap only, which occurs at the secular frequency ω_{sec}. The horizontal signal combines all motion in the horizontal coordinates of the trap; its first major axis in this plane points along the endcaps (z-coordinate), and the particle motion in this direction occurs at the frequency ω_z. The second axis in this plane is perpendicular to the z−axis, and the secular motion in this direction occurs at frequency ω_{sec}. The resulting damping voltages for the horizontal coordinates contain a mix of both damping signals. However, naturally a trapped particle reacts stronger to a properly (resonantly) phased damping voltage in each dimension, than it reacts to the potentially heating, but non-resonant and non-phase-matched other parts (e.g. noise or damping signals which are meant for the respective other coordinate). Hence, damping can be effective in all three spatial dimensions, although only two of them are measured, by the exploitation of known relations in the six-dimensional phase-space.

Some choices that were made in the concept shall be highlighted separately.

- The use of an ion gun for charging enables orders of magnitude larger charge-to-mass ratios [26] of the particles than other charging mechanisms that were used in previous concepts, such as charging via a discharge of the trap voltages [27]. The kinetic particle energy is related to its potential energy in the trap via $E_{kin}(0) = QU_p(x_{max})$. This shows, that the residual amplitude x_{max} for a given residual kinetic energy of the particle decreases with higher charge. The current system demonstrated the ability to consistently charge particles close to the physical limit of field ionization or dielectric breakdown.

- The most important parts of the concept are the optical setup and the optoelectronic damping which enable experiments with well positioned targets at high-vacuum conditions, such that we do not rely on buffer gas for particle positioning [27]. This is important to ensure that the high-power laser and diagnostics for the laser plasma interactions are not affected by a gaseous environment. Moreover, the setup allows variable implementations in various target chambers, e.g. in that the distance between both identical lenses is adjustable to fit specific chamber geometries.

- The system collects and evaluates small amounts of stray light scattered by tiny target particles. Since just the center-of-mass of the particle's magnified image is registered by the PSD, which does not necessarily need to be in perfect focus, it was decided to use the earlier mentioned large aperture f/1 lenses with a comparably small depth of focus ($\sim$1–2 μm). The potential disadvantage of limited depth of focus is easily compensated by the increased efficiency in light-collection. E.g. the high-numerical-aperture lenses allow to collect several nW of optical power from a 1 μm particle. For even smaller particles down to 100 nm an optional image-intensifier with a gain of 10^3 is available. This can be placed in the 1:1 image outside of the chamber and artificially increase the PSD signal-to-noise ratio.

The fiber-coupled laser is brought into the vacuum chamber through a vacuum feed-through. This—together with the light-collecting powermeter—serves to reduce stray-light sources other than that the target particle itself. It also allows positioning of the illuminating laser arbitrarily within the chamber in a compact adaptable way.

4.3   Selected Measurements of Relevant Single Particle Parameters

4.3.1   Charge-to-Mass Ratio

The real-time tracking of the particle enables the measurement of the secular frequency.[5] For this purpose, the PSD coordinate signal in vertical dimension was measured for a 12-second long period in the time-domain. The power spectral density of the recorded motion was calculated via discrete Fourier transformation of the time-domain signal and revealed the secular frequency as the dominant peak in the

---

[5]This section including figures is reproduced with permission from the original peer-reviewed article: T.M. Ostermayr et al., Review of Scientific Instruments, **89**:013302, (2018). The article is published by the American Institute of Physics and licensed under a Creative Commons Attribution 4.0 International License (https://creativecommons.org/licenses/by/4.0/).

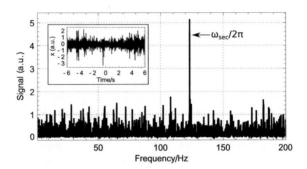

Fig. 4.4 Metrology of the particle secular motion. Example of a trajectory measurement in the time-domain (inset) for the vertical PSD signal, and corresponding power-spectral density (frequency measurement) via Fourier transformation of the time-domain signal. Figure and caption reproduced with permission from [21]

frequency domain. An example of such measurement is displayed in Fig. 4.4. During these measurements, the particle trajectory was adjusted to fill $\sim 3/4$ of the PSD field-of-view, corresponding to $\sim 300\,\mu m$ in the real space. All damping measures were turned off during the measurement. The peak could be unambiguously identified as the particle's secular frequency, because it shifted according to the expectations when the trap-parameters (Ω, V) were changed. From the secular frequency, the charge-to-mass ratio Q/M and the physical strength of the confinement in the trap in terms of the pseudopotential can be inferred (using the known relation of q and ω_{sec} given in the theory section, the definition of U_p in Eq. (4.7) and the known trap parameters r_0, Ω and V). Both are strong indicators for the ability of the trap to accurately position a particle. In the following example, we demonstrate the method for a polystyrene target of $10\,\mu m$ diameter. Figure 4.5 shows secular frequencies measured for six consecutively trapped particles with trap parameters $\Omega/2\pi = 1.1\,kHz$, $V = 1.0\,kV$ and r_0 as specified earlier. Error bars are inferred from the spectral width of the frequency peak and scaled up by a factor of 100 for better readability. The mean secular frequency is found as $\omega_{sec}/2\pi = 125\,Hz$, given as black solid line, with a standard deviation of 9 Hz (indicated by the gray band). The example reveals a reproducible charge-to-mass ratio of 0.29 C/kg corresponding to a surface potential of the sphere of 288 V. The corresponding electric field at the particle surface of 58 MV/m is in the range of the dielectric strength of the material [28] of about 20 MV/m, that is expected to limit the achievable particle charge [26]. At pressures of 10^{-6} mbar, the value stays constant for many hours [29]. The q-parameters are inferred close to 0.3. Consistency checks were performed by tuning q to the region where particle trajectories naturally become unstable ($q = 0.908$) by lowering the trap frequency. The particle trajectories consistently turned unstable and left the trap, for frequencies of $675\,Hz > \Omega/2\pi > 650\,Hz$, corresponding to $q \cong 0.908$.

From these measurements and the well defined particle mass, it is also straightforward to retrieve the charge carried by the specific target. For $\omega_{sec}/2\pi = 125\,Hz$ in the example, that is $Q = 9.9(0.9) \cdot 10^5$ e. The specified error stems from the distribution

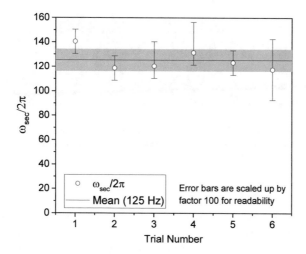

Fig. 4.5 Reproducibility of the charge-to-mass ratio. Secular frequencies for particles trapped consecutively, measured at $\Omega/2\pi = 1100\,\text{Hz}$. The horizontal line shows the mean value of $\omega_{sec}^{mean}/2\pi = 125\,\text{Hz}$ and the gray band indicates the standard deviation. Error bars are inferred from the spectral width of single measurements and scaled by a factor 100 for readability. These measurements reveal the comparably small variation of secular frequencies and charge-to-mass ratios, demonstrating the overall stability of the charging mechanism. Figure and caption reproduced with permission from [21]

of particle diameter in the employed sample, specified with 3% root-mean-square deviation. The trapped particle itself was comprised of $5 \cdot 10^{13}$ hydrogen and carbon atoms, easily outnumbering the number of elementary charges brought onto it. For laser-plasma experiments it is therefore valid to consider the trapped particle as unvaried from its original specification. The charge Q leads further to the potential energy $QU_p = 69\,\text{MeV}$ or $11\,\text{pJ}$ evaluated at the distance r_0 from the trap center. This is more than an order of magnitude larger than the particle energy resulting from a free fall from a height of several cm, and facilitates our straightforward trapping process in the first place. It shall be mentioned that this specific trap setup was not laid out or optimized by any means for high-resolution spectroscopic measurements. Still, the presented measurements and calculations serve to characterize particle and trap properties sufficiently well for our purpose. Thereby, a reliable, controlled, and stable trapping procedure was established as a basis for the trajectory damping and, ultimately, for laser-plasma experiments [21].

4.3.2 Damping Sequence and Residual Motion

After the initial trapping process of several tens of seconds, which involves the charging and trapping of the particle and the ramping of the endcap voltages, the electronic damping comes to full effect and reduces the amplitude of the particle

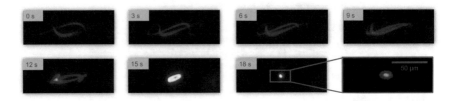

Fig. 4.6 Damping sequence. Typical damping sequence for a $2\,\mu m$ particle after ramping the endcap voltages and stopping the ion gun. The sequence was recorded with the overview camera with exposure time of 50 ms, while the last frame corresponds to the 20X magnified focus diagnostics image

trajectory. The complete duration of this process varies from several seconds to few minutes depending on the particle size. In Fig. 4.6 we show snapshots of the damping process for a $2\,\mu m$ polystyrene particle, where the temporal origin is set after the initial steps. The pressure in the chamber during the damping is $10^{-6}\,mbar$. In the example sequence, the particle trajectory is confined to its final precision within 18 s. After the initial cooling process, the particle can stay confined for hours and longer.

The most important measurement for the practical usability of trapped particles as targets is the quality of the damping, quantified by the amplitude of the residual motion of a trapped particle.[6] We measured the extent of the remaining particle motion using the emulated focus diagnostics. The 20-fold magnification allowed to resolve the motion. For Fig. 4.7 we trapped a 10-μm diameter polystyrene particle at 1500 V, 1400 Hz rod voltage and 250 V DC endcap voltage. We captured 12900 frames with 70 μs exposure time each, at a frame-rate of 7.5 Hz. The exposure time is shorter than micro- and macromotion oscillation periods of the particle. From each image, we determine the center of mass and plot the two-dimensional histogram of the occurrence of a specific particle position in Fig. 4.7. The Full-Width at Half Maximum (FWHM) of the motion is confined to 2.9 μm in horizontal and 1.9 μm in vertical direction. Once the particle is damped to such level of residual motion, it can stay confined for hours and longer. The white circle refers to the FWHM diameter of a typical high-power laser focus of 5 μm. From this, we can expect a hit-probability (defining hit with at least half-peak intensity interaction with the sphere center) of 94.5%. This estimate does not take into account pointing instabilities of typical large-scale high-power laser systems that are often of the order of their FWHM spot-size.

We note that the presented center-of-mass tracking is the most straightforward and accurate method available to determine residual motion, since more direct methods (e.g. time-integrated imaging) are easily flawed by the particle motion along the microscope dimension (out of the image plane), and by the complex optical

---

[6]From this sentence on, the section and Fig. 4.7 are reproduced with permission from the original peer-reviewed article: T.M. Ostermayr et al., 89:013302, Review of Scientific Instruments, (2018). The article is published by the American Institute of Physics and licensed under a Creative Commons Attribution 4.0 International License (https://creativecommons.org/licenses/by/4.0/).

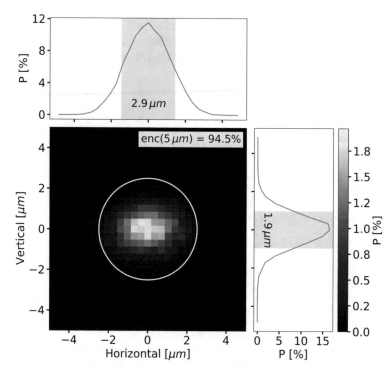

Fig. 4.7   Residual motion. Distribution of tracked center of mass for a 10-μm diameter trapped polystyrene sphere imaged at 20-fold magnification. The camera exposure time of 70 μs for each frame recorded at 7.5 Hz was chosen short enough to resolve single points of the particles micro motion. The convoluted hit probability of a 5 μm FWHM laser-focus would be 94.5%. In particular, positioning is more accurate than typical laser-pointing jitters. Figure and caption reproduced with permission from [21]

properties of transparent micro-particles, which can both lead to blurred images even for particles at rest.

The measurement in Fig. 4.7 was performed in our test-laboratory situated in the first floor of a laboratory and office building. Tables of the setup are built of solid aluminum structures and do not implement further damping techniques. It is notewor-thy that measurements during general day-time at which the building is comparably populated, are sensitively influenced by all kinds of vibrations caused by this pop-ulation. Presented measurements regarding damping and stability were recorded in less populated times. For technical reasons, the majority of large laser systems is situated in ground-level or basement floors, often taking special care for vibration stability. This comes to the advantage of our setup in its designated environment. In high-power laser-plasma experiments using our apparatus, the hit percentage of laser shots using 5–10 μm FWHM laser foci on target has constantly been larger than 50% and mostly limited by laser-pointing jitters. This reconfirms the usefulness of the target system for laser-microplasma investigations. Spherical polymer particles with diameters ranging from 500 nm to 50 μm as well as tungsten particles with diameters up to 10 μm have been trapped and positioned with comparable precision using this setup, and used in experiments presented in this thesis.

References

1. Hilz PB (2009) Diploma thesis: target developments for laser plasma experiments. MA thesis, Munich
2. Ostermayr T (2012) Mass limited targets for laser driven ion acceleration. MA thesis, LMU
3. Haffa D (2014) Optimisation and application of mass limited levitating targets in laser plasma experiments. MA thesis, LMU, München
4. Singer M (2015) Levitating nano graphene platelets for laser plasma experiments. MA thesis, LMU
5. Gebhard J (2016) Laser ion acceleration using reduced dimension targets. MA thesis, LMU, München
6. Major FG, Gheorghe VN, Werth G (2005) Charged particle traps, vol 37. In: Springer series on atomic, optical, and plasma physics. Springer, Heidelberg
7. Werth G, Gheorghe VN, Major FG (2009) Charged particle traps II, vol 54. In: Springer series on atomic, optical, and plasma physics. Springer, Heidelberg
8. Earnshaw S (1842) On the nature of the molecular forces which regulate the constitution of the luminiferous ether. Trans Camb Philos Soc 7:97–112
9. Meixner J, Schäfke FW (1954) Mathieusche Funktionen und Sphäroidfunktionen, vol LXXI, 1st edn. Die Grundlehren der Mathematischen Wissenschaften in Einzeldarstellungen, Springer, Heidelberg
10. Dehmelt HG (1968) Radiofrequency spectroscopy of stored ions I: storage part II: spectroscopy is now scheduled to appear in volume V of this series. In: Bates DR, Estermann I (eds) Advances in atomic and molecular physics supplement C, vol 3. Academic Press, pp 53–72
11. Stenholm S (1986) Semiclassical theory of laser cooling. Rev Mod Phys 58:699–739
12. Wuerker RF, Shelton H, Langmuir RV (1959) Electrodynamic containment of charged particles. J Appl Phys 30(3):342–349
13. Denison DR (1971) Operating parameters of a quadrupole in a grounded cylindrical housing. J Vac Sci Technol 8(1):266–269
14. Dayton IE, Shoemaker FC, Mozley RF (1954) The measurement of two dimensional fields. Part II: study of a quadrupole magnet. Rev Sci Instrum 25(5):485–489
15. Lee-Whiting GE, Yamazaki L (1971) Semi-analytical calculations for circular quadrupoles. Nucl Instrum Methods 94(2):319–332
16. Blaum K et al (1998) Properties and performance of a quadrupole mass filter used for resonance ionization mass spectrometry. Int J Mass Spectrom 181(1–3):67–87
17. Wang Y, Franzen J, Wanczek KP (1993) The non-linear resonance ion trap. Part 2. A general theoretical analysis. Int J Mass Spectrom Ion Process 124(2):125–144
18. Vedel M et al (1998) Evidence of radial-axial motion couplings in an rf stored ion cloud. Appl Phys B 66(2):191–196
19. Alheit R et al (1995) Observation of instabilities in a Paul trap with higher-order anharmonicities. Appl Phys B 61(3):277–283
20. Pedregosa J et al (2010) Anharmonic contributions in real RF linear quadrupole traps. Int J Mass Spectrom 290:100–105
21. Ostermayr TM et al (2018) A transportable Paul-trap for levitation and accurate positioning of micron-scale particles in vacuum for laser-plasma experiments. Rev Sci Instrum 89:013302
22. CGC Instruments, Chemnitz, Germany
23. Microparticles GmbH, Berlin. http://www.microparticles-shop.de. Retrieved 30 June 2017

24. Tectra GmbH. www.tectra.de. Retrieved 25 Feb 2016
25. Sitek, PSD Model: 2L10-SU7, http://www.sitek.se/pdf/psd/S2-0003-2L10_SU7.pdf/. Last accessed 12 Oct 2017
26. Cermak I (1994) Laboruntersuchung elektrischer Auadung kleiner Staubteilchen. PhD thesis, Ruprecht-Krals-Univeristät, Heidelberg
27. Sokollik T et al (2010) Laser-driven ion acceleration using isolated mass-limited spheres. New J Phys 12(11):113013
28. Harper CA (2000) Modern plastics handbook. McGraw-Hill
29. Pavlů J et al (2007) Interaction between single dust grains and ions or electrons: laboratory measurements and their consequences for the dust dynamics. In: Faraday discussions, vol 137, p 139

Part III
Laser-Microplasma Interactions

Chapter 5
Laser-Driven Ion Acceleration Using Truly Isolated Micro-sphere Targets

While laser-plasma interactions at lower intensities with spherical targets are reasonably well understood (e.g. [1, 2]), only few experimental approaches have been made at laser-intensities of $a_0 \gg 1$ to study their dynamics. As pointed out in the introduction, approaches were performed with cluster-targets in the 100 nm range and droplet targets in the several μm range. Such targets bring a large number of atoms into the laser-focus aside from the actual target, and shallow density gradients at the target-vacuum boundary are inevitable. It has been shown just recently, that such density gradients around the target give rise to Weibel-type instabilities and influence laser-ion acceleration [3, 4]. Our approach is to use targets of initially solid material, like plastics, and levitate those in focus with the Paul trap apparatus. Due to the limitations of previous target-mechanisms, neither target-size nor laser intensity have been varied in sufficient ranges (even more so in a single experiment) to truly explore different acceleration mechanisms.

This chapter aims to theoretically, numerically and experimentally explore the realm $a_0 \gg 1$ of spherical plasma explosions in a broad parameter-range. In the first step we introduce a theoretical framework for such system. This covers mechanisms of ion acceleration in single and multi-ion-species targets. Predictions based on laser and target parameters are developed for the occurrence of different acceleration processes.

In the second and third steps of this chapter, particle-in-cell simulations and experiments provide a refined image of the laser-plasma interaction and confirm the theoretical approach.

In the last step, the knowledge from the theoretical analysis is leveraged to intentionally enter an advanced acceleration mechanism showing highly directional acceleration of quasi-monoenergetic protons from a spherical target.

© Springer Nature Switzerland AG 2019
T. Ostermayr, *Relativistically Intense Laser–Microplasma Interactions*,
Springer Theses, https://doi.org/10.1007/978-3-030-22208-6_5

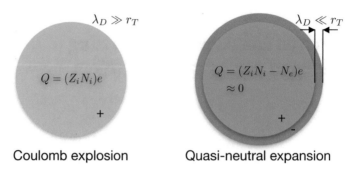

Fig. 5.1 Schematic of the considered initial situation for Coulomb-explosion and quasi-neutral expansion. Q denotes the net macroscopic charge-buildup caused by the heated electrons

5.1   Theory

Before going into more depth, an overview of distinct acceleration mechanisms for mass limited targets of spherical shape is given. Naturally, some of those are closely related to respective plane foil-scenarios.

Most analytical models consider the plasma-vacuum interface in a single dimension, quite similar to what has been introduced earlier in the discussion of Debye shielding. Usual analytical models of the laser ion-acceleration do not consider the laser-plasma interaction itself. Instead, they start off with electron- and ion-distributions derived via scalings from the target- and laser-parameters (e.g. [5–8]). In addition, they are typically constrained by the postulated symmetry (e.g. planar or spherical). In the following we discuss the Coulomb explosion of a sphere from which all electrons have been stripped in one limiting scenario, and the spherical expansion of a quasi-neutral (i.e. cold) sphere as the other limit (cf. Fig. 5.1). The range in between these extremes, evinces mixed characteristics.

A comprehensive way to cover the broad range of dynamics in spherical expansion including the interesting intermediate parameter-ranges in a single framework, was introduced by [8]. Using a one dimensional spherical gridless particle code, the method reproduces predictions of several analytical models by numerical calculation and allows to investigate the range from Coulomb explosion to quasi-neutral expansion, for single and multi-species targets of arbitrary initial velocity and density distributions, without losing the simplicity of a one dimensional and spherically symmetric model. We reproduced the code for quick calculations of expectations for a given set of target parameters and electron temperatures, calculated from the laser parameters assuming the applicability of scaling laws (e.g. via Èq. (2.45)).

In the second part of this section we discuss effects caused by multi-ion-species targets based on the same numerical method.

Last but not least, we consider the relevance of each acceleration mechanism for given laser and target parameters. In contrast to earlier work , the dynamic change of laser and target-parameters during the interaction with the laser pulse is identified as a crucial issue for predictions and interpretations of multidimensional particle-in-cell simulations and experiments. In this framework, a simple theoretical approach is developed to include these effects.

5.1.1 Characteristic Ion-Spectra of Different Mechanisms

Taking the approach from an experimentalist point of view, we focus on time-integrated characteristics such as ion kinetic energy distribution and maximum ion kinetic energy for specific scenarios, which can be compared to numerical simulations and experimental data.

Coulomb Explosion

In an ideal Coulomb explosion, electrons do not play a role in the ion acceleration except for being absent. Ions are accelerated in radial symmetry by the repulsion of the positive charge-surplus inside the target, created by the ions themselves. For the Coulomb explosion of a uniformly charged sphere of radius r_T containing a single species of ions, ion kinetic energy distributions are accessible analytically. Initially, the electric field of the charged spherical target at position r from the sphere center is given via Gauss' theorem as

$$\mathcal{E}_f(r) = \frac{Qr}{4\pi\varepsilon_0 r_T^3}, \tag{5.1}$$

for $r < r_T$, and $Q = 4/3\pi r_T^3 n_i Z_i e$ representing the sphere charge respectively. In this context Z_i and n_i stand for charge state and the number density of the ion species inside the target. For $r > r_T$ the electric field due to the charged sphere drops with r as

$$\mathcal{E}_f(r) = \frac{Q}{4\pi\varepsilon_0 r^2}, \tag{5.2}$$

similar to the field of a point-charge Q situated at the origin. Obviously, $\mathcal{E}_f(r)$ is maximized for $r = r_T$, meaning that ions starting in that outermost position will be accelerated to higher energies than particles further inside the target, and more precisely $\mathcal{E}_f$ is a monotonically increasing function of r in the volume occupied by charged particles. Particles starting further out will not be overtaken by ones starting inside the sphere [8]. Using this, we can consider any charged particle with $q = Z_i e$ inside the sphere at position $r < r_T$ as being driven only by charges initially located inside the interval $[0, r]$. The kinetic energy gained by a particle starting at that position can be found by integration of Eq. (5.2) considering only the charge included within this interval, $Q_{[0,r]} = 4/3\pi r^3 n_i Z_i e$

$$E(r) = e\left(\phi(\infty) - \phi(r)\right) = \frac{1}{3\varepsilon_0} r^2 n_i Z_i^2 e^2. \tag{5.3}$$

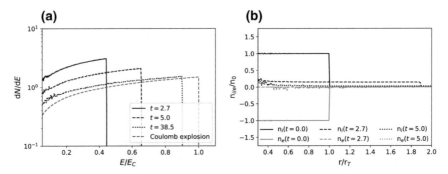

Fig. 5.2 Coulomb explosion. Temporal evolution of **a** the (normalized) proton spectrum and **b** the plasma density in a Coulomb explosion with time, with $k_B T_e = 2\, E_C$. Times are specified in units of $\omega_p t$. The red line in **a** shows the analytical spectrum from Eq. (5.6) for a pure Coulomb explosion. Simulated data were obtained with an adaption of the gridless particle code from [9]

Again, $E(r)$ is a monotonically increasing function of r and the maximum energy will be gained by particles starting at the outermost position of the sphere at $r = r_T$ reaching the potential of the fully charged sphere

$$E_C = \frac{1}{3\varepsilon_0} r_T^2 n_i Z_i^2 e^2. \tag{5.4}$$

The spectral distribution dN/dE can be deduced using the infinitesimal particle number element $dN = n_i dV = n_i 4\pi r^2 dr$ and

$$\frac{dE}{dr} = \frac{2}{3\varepsilon_0} r n_i Z_i^2 e^2, \tag{5.5}$$

from which the spectral form follows from $dN/dr = dN/dE \cdot dE/dr$

$$\frac{dN}{dE} = \frac{6\pi\varepsilon_0}{Z_i^2 e^2} r(E) = \frac{\sqrt{108}\pi\varepsilon_0^{3/2}}{n_i^{1/2} Z_i^3 e^3} \sqrt{E}. \tag{5.6}$$

As observed in Fig. 5.2 the spectrum characteristically increases as $E^{1/2}$, up to the maximum energy, where the sharp initial target-vacuum boundary causes the cutoff. The initially uniform density distribution is maintained throughout the explosion.

Interestingly, the simplicity of the problem immediately breaks down when considering even small gradients in the initial density distribution. That is due to wave-breaking effects, i.e. particles overtaking each other. For similar reasons, multi-species targets constitute a more difficult scenario. Such problems have been addressed via kinetic approaches in a number of works including [9–12].

Fig. 5.3 Quasi-neutral expansion. Analytical quasi-neutral expansion spectrum at $k_B T_e = 0.005$, 0.01 and 0.05 E_C. All spectra are normalized with respect to their integral

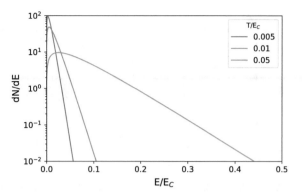

Quasi-Neutral Expansion

In the very low temperature limit quasi-neutrality ($n_e = Zn_i$) is implied except for scale-length smaller than the Debye-length, which itself is considered small with respect to the target-size for this mechanism. For such a system with spherical symmetry, solutions have been found in the hydrodynamic approximation [5] (isothermal and adiabatic), directly [6, 7] (adiabatic) and in numerical or semi-numerical calculations [8, 13] (isothermal and adiabatic). These solutions show a universal ion spectral distribution for $t \to \infty$, because the ion density distribution approaches the same profile over time, independently from the initial ion-density distribution [5, 8]. This spectral distribution

$$\frac{\mathrm{d}N}{\mathrm{d}E} = \frac{2\sqrt{E}}{\sqrt{\pi(k_B T_e)^3}} \exp(-E/k_B T_e). \tag{5.7}$$

is the well known Maxwell energy distribution, which is naturally assumed with time by an homogeneously heated mass approaching thermal equilibrium.

This distribution extends to infinity and does not produce a high-energy cutoff that could be denoted as the maximum ion energy. However, particle numbers show a very steep (exponential) decay towards high energies, defined by the small temperatures required for the quasi-neutral approximation to be valid. Examples of such spectrum are shown in Fig. 5.3. The validity of this spectral distribution breaks already at finite characteristic (thermal) energies, e.g. with Debye lengths approaching the percent range of the target diameter. The effects thereof will be discussed in the following.

Ambipolar Expansion

The intermediate regime of 'ambipolar expansion' occurs whenever the electron temperature is small enough to keep electrons in the vicinity of the plasma, while being hot enough to break quasi-neutrality, e.g. with the Debye length approaching the plasma-size. This realm is characterized by a mixture of characteristics found in quasi-neutral and in Coulomb explosion. Even for very small finite electron temperatures (and hence small Debye lengths) the ion-spectrum features a clear

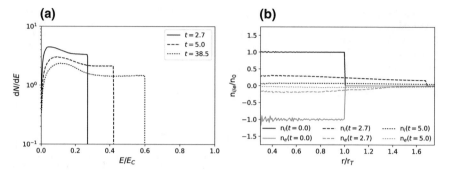

Fig. 5.4 Ambipolar expansion. Temporal evolution of **a** the (normalized) proton spectrum and **b** the plasma density in an ambipolar expansion with time with $T_e = 0.2\ E_C$. Times are specified in units of $\omega_p t$. Obtained with an adaption of the gridless particle code from [9]

maximum cutoff energy instead of an infinitely decaying spectrum towards higher kinetic energy. Again, the cutoff represents the initially steep target-vacuum boundary expanding into the vacuum. For growing electron temperature the ion spectral distributions slowly transform from the quasi-neutral to the Coulomb explosion shape, while the relative cutoff energy (i.e. relative to the Coulomb energy E_C) increases and reaches its maximum in the Coulomb explosion case. In ambipolar scenarios, some electrons remain bound to the target and buffer the Coulomb field reducing the ion-energy as compared to the Coulomb explosion energy for a given target. The analytical treatment of the transition via self-similarity (non-uniform initial and final electron and ion density and velocity distributions) in an adiabatic system [14] can give some insight: the solution in this special case is indeed a simple linear combination of the separate spectral distributions for quasi-neutral and Coulomb explosion cases discussed above. The spectral distribution for the initially uniform target, as exclusively discussed here, differs from this analytical solution quantitatively and is not analytically accessible yet. Still, it shares the qualitative features of the 'combined' spectrum, as observed in Fig. 5.4.

Figure 5.5 summarizes the spectral shape in the limiting cases of Coulomb explosion and quasi-neutral expansion, and for the ambipolar expansion in between, for initially uniform spheres. The transition across acceleration mechanisms is continuous in terms spectral shape and maximum energy. Obviously, for a given target size and density, the maximum energy attainable in spherically symmetric explosions is that gained in the Coulomb explosion. A good handle on the maximum energy scale E_{max} when scanning the electron temperature through the acceleration mechanisms can be obtained from the aforementioned analytical solution in Ref. [14] as

$$E_{max} = 2k_B T_e W\left(\frac{r_T^2}{2\lambda_D^2}\right).$$
(5.8)

Here W is the inverse function of $x = W \exp W$ known as Lambert W function [14]. Numerical calculations show that this solution is still a good approximation, even

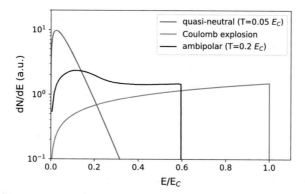

Fig. 5.5 Summary of acceleration mechanisms. Analytical quasi-neutral expansion spectrum at $k_B T_e = 0.05 E_C$, ambipolar expansion spectrum at $k_B T_e = 0.2 E_C$ for a uniform sphere obtained with the numerical code, and analytical Coulomb explosion spectrum for infinite temperature. All spectra are normalized with respect to their integral

when considering the uniform target. Asymptotically, the solution recovers Eq. (5.4) for the Coulomb explosion case $(r_T/\lambda_D \ll 1)$ with $W(x) \approx x$. In the opposite limit $(r_T/\lambda_D \gg 1)$ the scaling may be used with $W(x) = \ln(x/\ln x)$ [14].

5.1.2 Multi-species Plasmas and Target Size-Scan

In our simulations and experiments we employed targets consisting to equal parts of homogeneously distributed carbon atoms and hydrogen, both considered to be fully ionized. The interaction in between the different ion-species (sometimes referred to as a shock) contributes to the overall interaction and significantly changes the ion kinetic energy distribution.

In addition, in realistic laser-plasmas the laser-intensity can not be scaled easily to span many orders of magnitude (e.g. such that $k_B T_e$ ranges from 0.01 E_C to 10 E_C). In order to observe varying characteristics of acceleration mechanisms in such a broad range, we rely instead on a target-size variation (thus varying E_C) at constant laser parameters.

A first glance at such parameter-scan can be obtained from the gridless particle code [9]. For a given laser-intensity we consider all electrons to be instantaneously heated to the pondermotive potential, according to Eq. (2.45), while the system expands adiabatically thereafter (i.e. no further energy is added to the system). In Fig. 5.6, ion kinetic energy distributions are presented for the electron temperature of 4.3 MeV and electron density of $3.4 \cdot 10^{23}$ cm^{-3}. While the maximum kinetic energy and the overall particle count for either ion species drops with target-size as expected, a qualitative change in the proton kinetic energy spectra is observed. Starting with Fig. 5.6a, the proton-spectrum features two energetically separated peaks, where the

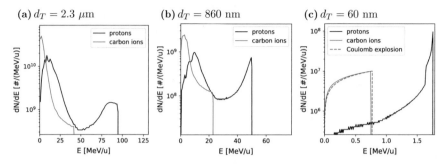

Fig. 5.6 Multi-ion-species expansion. Ion kinetic energy spectrum obtained from initially uniform multi-ion-species target (C^{6+} and H^+) with electron density of $3.4 \cdot 10^{23}$ cm^{-3} (corresponding to solid density polystyrene) scaled to an electron temperature of $k_B T_e = 4.3$ MeV. Calculated with the spherical gridless particle code [9]. The red dashed line in **c** shows the analytical spectrum Eq. (5.6) for a Coulomb exploding carbon sphere, disregarding any effect caused by the protons contained in the target

low-energy peak clearly dominates the high-energy peak in terms of particle count. The carbon spectrum corresponds to that of a cold ambipolar plasma-expansion. The maximum velocity of carbons corresponds to about half the maximum proton velocity due to its charge-to-mass ratio being a factor of two smaller. With the smaller spheres in Fig. 5.6b and c, the qualitative behavior shifts; the high-energy peak narrows and grows in relative particle number, while the low-energy peak decreases and even vanishes, the closer the system approaches Coulomb explosion conditions. This can be understood, since the carbon charge increasingly efficient pushes protons out of the volume occupied by carbon-ions. Meanwhile, the carbon-ion behavior in Fig. 5.6c remains almost undisturbed from the single-species scenario of a Coulomb explosion in Fig. 5.5. Qualitatively, these results are intuitively understandable. The maximum energy scales with Eq. (5.8) and shows decreasing energy towards decreasing target diameter. However, the validity of underlying assumptions must be challenged; that is particularly the assumption regarding homogeneous, isotropic and instantaneous electron heating to a certain energy. Such considerations will be subject to the following sections. At this point it shall be highlighted, that the clear quantitative and qualitative changes in signatures of spectral ion distributions do suggest their application to identify relevant mechanisms in experiments and simulations.

5.1.3 Influence of Laser and Target Parameters

After introducing these basic expectations for spectral distributions of ions, this section aims to quantitatively determine the relevant acceleration mechanism for given experimental conditions, including to some extent the laser-plasma interaction itself.

From Coulomb Explosion to Ambipolar Expansion

We can make two basic estimates for when Coulomb explosion will become significant. In a first line of thought, the Coulomb explosion will start to dominate if the mean electron kinetic energy surpasses the Coulomb potential of the fully charged sphere Eq. (5.4)

$$k_B T_e > E_C. \tag{5.9}$$

This condition is equivalent to

$$\sqrt{3}\lambda_D > r_T, \tag{5.10}$$

where the Debye length is modified by a geometrical factor for the spherical geometry. As this condition describes the transition to Coulomb explosion we can use the ponderomotive scaling for the electron temperature of free electrons. I.e. instead of using the cold-plasma approximation for the cycle averaged Lorentz-factor in Eq. (2.45), we use the ponderomotive potential for the free electron Eq. (2.32) following Ref. [15]. Inserting Eq. (2.32) to either of the equivalent conditions Eqs. (5.9), (5.10) leads to

$$a_0 > \frac{4\pi}{\sqrt{3}} \sqrt{\frac{n_e}{n_c}} \frac{r_T}{\lambda}. \tag{5.11}$$

Condition Eq. (5.11) describes the strength of the laser with wavelength λ, which would allow to instantaneously and fully remove the *last* electron from an ionic sphere ($n_e = 0$, $Z_i n_i = n_{e0}$) of radius r_T [15]. Using Eq. (2.46) with laser energy incident on the target $E_L = I \pi r_T^2 \tau_L$, and pulse duration τ_L, the hot-electron number N_h determined from Eq. (5.11) is

$$N_h = 2\pi r_T^2 \tau_L c n_c \eta, \tag{5.12}$$

where η is the fraction of absorbed laser energy, usually set to 0.5. Interestingly, the number of heated electrons N_h does not relate to a_0 but only to the effective laser-volume and the critical density. An interesting quantity is the relation of this number of heated electrons to the number of electrons contained in the target N_T

$$\frac{N_h}{N_T} = \frac{3}{2} \frac{n_c \eta}{n_e} \frac{c \tau_L}{r_T} = \frac{2 V_L n_c \eta}{V_T n_e}, \tag{5.13}$$

where V_L is the effective volume of the laser and V_T is the volume of the target. We will continue to observe this ratio as we introduce further constraints.

In a second line of thought, in close analogy to the derivation of optimum conditions for radiation pressure acceleration for foil-targets [16, 17], we can relate the radiation pressure of the laser with the electrostatic pressure caused by *all* electrons of the sphere being pushed out of the sphere at the same time (but not to infinity) [18, 19]. With such pressure balance, the ions inside the target could readily start to evince effects caused by the Coulomb repulsion, e.g. between different ion-species. The pressure (in)-balance yields

$$\epsilon_0 \mathcal{E}_C^2 \leq 2\frac{I}{c}, \tag{5.14}$$

where $\mathcal{E}_C = \mathcal{E}_f(r_T)$ is the Coulomb-field of the fully charged sphere at its surface Eq. (5.2). The required normalized field strength for the laser is

$$a_0 \geq \frac{2\pi n_e}{3n_c}\frac{r_T}{\lambda}. \tag{5.15}$$

Condition Eq. (5.15) is very similar to the corresponding result for flat targets ($a_0 \geq n_e l_T/n_c \lambda$, with target thickness l_T), except for some constant factors. Again, using Eq. (2.46) to determine the ratio of hot electron number to the number of electrons contained in the target yields

$$\frac{N_h}{N_T} = \frac{\eta}{2\sqrt{2}}\frac{c\tau_L}{\lambda}. \tag{5.16}$$

Note how both Eqs. (5.16) and (5.13) depend on the geometrical length of the pulse $c\tau_L$ in relation to the target radius and the laser wavelength respectively. Both can therefore lead to the contradiction that more electrons are being accelerated (N_h), than are available in the target (N_T).

This brings us to the ambipolar expansion. This mechanism is then most efficient and delivers the highest ion kinetic energy when

$$\frac{N_h}{N_T} \stackrel{!}{=} 1, \tag{5.17}$$

while Coulomb conditions are not yet met. Using Eq. (2.46), the corresponding condition for the laser is

$$a_0 = \frac{4\sqrt{2}\cdot n_e}{3n_c}\frac{r_T}{\eta c\tau_L}, \tag{5.18}$$

where we have assumed that $a_0 \gg 1$ is still valid (in agreement with our experiments) and thus the hot electron temperature from Eq. (2.45) can be approximated as

$$k_B T_e \approx m_e c^2 \frac{a_0}{\sqrt{2}}. \tag{5.19}$$

It comes to no surprise, that once again the ratios of volumes V_T/V_L and densities n_e/n_c play the critical role; simply put the sphere can take larger intensities at larger radius and/or higher densities before running into suboptimal conditions for acceleration in the ambipolar regime. In the same sense, longer pulse durations require reduced intensities, to keep the total amount of incident energy constant and to stay in the optimum regime of acceleration.

Towards larger targets (or smaller intensities), such that $N_h < N_T$, the large number of unheated (cold) electrons can efficiently redistribute the hot electron energy among themselves, leading to an overall reduced temperature and similarly reduced ion kinetic energies. This effect quickly cancels the larger ion kinetic energy expected from larger targets according to Eq. (5.8) because its scaling with radius is stronger.

Towards smaller targets (or higher intensities), the situation is more complex. From Eqs. (5.13), (5.16) and (5.17) it is apparent, that a region in between the ambipolar expansion and the Coulomb explosion can exist, where neither of both is optimized, i.e. $N_h > N_T$ while none of the Coulomb conditions is met. We refer to this as the 'beyond optimum ambipolar expansion' in the following. It is worth to interpret the observation.

1. The first interpretation postulates the existence of a quasi-coherent electron-recycling mechanism that transmits the energy from electrons to ions instantaneously, such as in the simplified explanation of radiation pressure acceleration in planar targets [17]. In case of Eqs. (5.15)–(5.16), each electron would then continuously gain the ponderomotive potential per laser-cycle. This energy is redistributed efficiently and instantaneously to the ions, such that electrons always remain available as quasi-cold electrons. This continuous and instantaneous transfer of energy is equivalent to the suppression of electron heating. It thereby prohibits the classical Coulomb explosion (Eqs. (5.11)–(5.13)) and the classical ambipolar expansion (5.17), (5.18), as N_h is effectively zero. Instead, such mechanism shall be investigated with respect to Eq. (5.15) and potentially supports the directional acceleration of ions in schemes like radiation-pressure acceleration [17], 'solid bunch acceleration' [20], or 'directed Coulomb explosion' [21], where the laser imprints its directionality via the electrons to the ion beam, which is no longer isotropic.
2. If there was no quasi-coherent energy-transmission to ions, the evolution of the electron energy distribution will play a role and must be discussed. Two fundamentally different scenarios can be identified and are considered in this and in the following point. The first possibility is, that electron temperature continues to grow indefinitely as long as the laser is on, and finally exceeds the ponderomotive potential by a factor N_h/N_T, e.g. as given by Eq. (5.13) or Eq. (5.16). Multiply heated electrons with increased temperatures have been observed in simulations using isolated targets with comparably long laser pulses [22].
3. The second scenario in absence of a quasi-coherent energy transmission to ions postulates, that electron-heating will be limited by the nature of the interaction with the laser. E.g. the laser may heat the electrons to the ponderomotive potential of free electrons Eq. (2.32) instead of the ponderomotive scaling for a cold plasma Eq. (2.45), but not much further. This is likely, without special acceleration mechanisms at work. The only way the system could then react to the growing energy content, is a reduced laser-absorption efficiency η (i.e. smaller than the usually assumed 0.5–1) and hence reduced overall efficiency. In fact, even the multiply heated electrons observed by [22] did not show temperatures far beyond the ponderomotive scaling.

Which exact interpretation is suitable for a given experimental situation, needs to be clarified individually via numerical simulations and/or experiments. In the remainder of this work, we find the third scenario as the most realistic for our experimental and simulated situations. Note that these considerations are not arbitrary: they are implemented as per design in Eqs. (5.11), (5.15) and (5.18) via the choice of the most suitable electron temperature scaling with laser intensity in a given scenario.

Wrapping up this discussion based on the above argument, we can readily state, that the plasma response beyond $N_T = N_h$ may not behave as trivial as suggested in Fig. 5.6. It is expected to be directly affected by the laser-target interaction and its dynamics. The electron heating has an important impact on the energy scales expected from the ion acceleration. By reducing the target size such that $N_h > N_T$, the *rising electron temperature* would lead to a *higher ion kinetic energy*. Meanwhile, *smaller targets* generally lead to *smaller ion kinetic energy*. Both effects are counteracting and scale independently with the target radius; giving room for both to dominate in separate regimes. This hypothesis will reappear later in this thesis.

Transparency

The next consideration of relevance is, when the target can be considered as transparent. The relativistic transparency starts to play a role, when the plasma frequency matches the laser frequency

$$\omega_p^{rel} = \omega_L \tag{5.20}$$

where ω_p^{rel} is defined as earlier in this thesis, with the relativistic correction to the electron mass (i.e. $\omega_p^{rel} = \omega_p(a_0)$). Defined as such, the relativistic electron motion may result in induced transparency.

Besides the relativistic electron motion, an additional cause for induced transparency is the (three-dimensional) expansion of the target that takes place already during the laser-interaction, and the related reduction in electron density. While this effect is largely ignored for planar targets, it is of utmost importance for the spherical case relevant here. A target with density $n_e = 340n_c$ needs to expand just 7-fold in radius, to become classically undercritical/transparent.

The transparency of a target changes its interaction with the laser fundamentally. First of all, it realistically enables to access and heat all target electrons. Meanwhile, electron heating mechanisms that critically rely on overcritical plasma regions (Sect. 2.7) will be suppressed. Other mechanisms like the direct ponderomotive scattering of plasma electrons can take over: the entire target-volume can be accessed and all electrons can be heated by the transmitting laser pulse, to potentially higher temperatures (as required and implemented in Eqs. (5.11) and (5.15)). The induced transparency can therefore play a major role in the transition of acceleration mechanisms. In the transition to relativistic transparency, experiments carried out with foil-targets have demonstrated exciting results showing enhanced ion energies [23–25] and generation of (mono-)energetic electrons breaking out of the target [26].

Directed Acceleration

Beyond the pure spherical expansion of targets, it was already mentioned in context with the radiation pressure balance Eq. (5.15), that driving lasers in our experiments usually break the spherical symmetry. Several experiments have claimed the emission of directed ion-beams from laser-sphere interactions [21, 27]. The existence of electric fields that are non-spherical has been directly observed in laser-sphere interactions [28] via proton deflectometry. In certain scenarios, the pre-curvature of the target could account for the spatially Gaussian or Airy-intensity distribution of the laser [27] and support directed acceleration overcoming the non-coherent expansion described earlier in this chapter. If a coherent mechanism exists to accelerate ions in the laser direction, the interesting limit of such scenario is the equivalence of laser-energy and target-rest-mass

$$E_L = m_T c^2, \tag{5.21}$$

which might enable acceleration of the entire target to a speed close to the speed of light. Such a mechanism has been predicted via simulations [20] and is here referred to as 'solid-bunch acceleration'. Already soon after the invention of the laser, this constraint has been discussed in the context of the laser's potential to drive interstellar space propulsion [29–31]. In fact, just recently this idea gained renewed interest and is now actively pursued [32]. Since none of the above cited experimental work used truly isolated mass-limited targets, the true solid-bunch acceleration with the complete target being accelerated, stayed inaccessible to experiments thus far. For not using truly isolated targets, such results are prone to misinterpret effects caused by the more complex experimental situation.

As an intermediate result, Fig. 5.7 summarizes all of the above estimates for acceleration regimes, considering negligible pulse durations (single optical cycle at 1 μm, equal to 3.3 fs, where required). This shows a rather ordered transition from ambipolar expansion to Coulomb explosion and finally to relativistic transparency. The partition of the parameter space is consistent with widespread accepted theories for the transitions between these regimes (e.g. [18, 33, 34]). The simplicity arises because crossings between parameter boundaries occur only at intensities beyond $a_0 = 100$, and three of our constraints (Eqs. (5.11), (5.15) and (5.18)) are linear in r_T and do not cross at all.

Note the following though; a single cycle pulse that is intense enough to heat electrons beyond either of the Coulomb conditions Eqs. (5.11) and (5.15) (solid and dashed line in Fig. 5.7 respectively), does not even contain sufficient energy to promote all target electrons to become sufficiently hot electrons (dotted line). A pulse of larger intensity and single-cycle duration, or a pulse of similar intensity at longer pulse-duration (both leading to higher energy content) is thus required to enter the Coulomb explosion.

Similarly, in the single-cycle consideration a laser fulfilling $N_h = N_T$ will lead to Coulomb explosion instead of ambipolar expansion in contrast to previous discussions. Likewise, the radiation pressure balance or directed acceleration would not

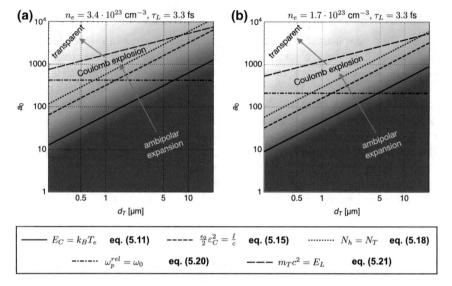

Fig. 5.7 Expansion regimes. Laser-ion acceleration mechanisms for isolated spheres with specified densities. The pulse duration is taken as a single optical cycle at wavelength 1 μm, so as to emulate the instantaneous nature of the heating process. Expansion during the pulse duration is neglected. Wherever needed (Eqs. 5.17 and 5.21), we assumed $\tau_L = 3.3$ fs corresponding to a single optical cycle at 1 μm, to determine the pulse energy. Regimes are given as colored areas and lines separating regimes by intensity are explained in the legend. Simplification Eq. (5.19) was not used here

be accessible in this parameter space. However, the single optical cycle does neither correspond to the typical laser used in laser-ion acceleration, nor to the lasers used in this work. In the following sections, we consider the effects of a finite pulse duration on the laser-plasma interaction.

Pulse Duration—Isothermal and Adiabatic Acceleration

The observation that even moderate sphere expansion during the pulse could lead to induced transparency due to the reduced density, and the observation that a certain pulse duration is required in order to have enough energy contained in the laser pulse to truly reach into Coulomb explosion (cf. previous section), raises the important question: which are the effects of target-expansion during the laser-irradiation? Thus far we have discussed the spherical target without taking into account such dynamics during the laser-plasma interaction. The pulse duration was exclusively used to consider the effective volume of the laser. In reality, the target density and radius are subject to change, due to the ongoing expansion, already during the laser-target-interaction.

To estimate the effect, the laser-pulse duration can be related to the timescale of plasma-expansion, $\delta = \tau_L / (d_T / c_s)$. Here $c_s = (Z_i k_B T_e / m_i)^{0.5}$ is the ion sound velocity in the plasma with m_i representing the ion mass.[1] Experiments using $\delta \sim$ 0 can be considered as **adiabatically** expanding plasmas that were excited by a delta-peak, resembling the simple calculations used above, and left alone afterwards. Values of $\delta \gg 1$ mark **quasi-isothermal** processes, where energy is continuously fed into the system at a constant rate during the expansion. Many experiments, including the ones presented within this thesis, operate around $\delta \sim 1$ and thus could be regarded as a combination of adiabatic and isothermal processes.

In these cases it is necessary to model the target-expansion during the laser-pulse interaction. Before considering more difficult temporal intensity distributions, we consider a temporal flat-top profile with instant rise-time and defined pulse duration τ_L. In order for the estimates introduced earlier to still be valid in a finite-pulse regime, we account for the changed dynamics in replacing r_T by $r_T = r_T(t = 0) + \tau_L c_s$ wherever suitable in the inequalities. In the limit of a single optical cycle (Fig. 5.8a and b), the parameter-space resembles the non-expanded scenario Fig. 5.7. For longer pulses, the expansion affects the inequalities in different ways (cf. Fig. 5.8c and d), thereby creating a much richer set of possible acceleration mechanisms that are experimentally accessible. Targets will evolve dynamically throughout their interaction with the laser pulse, and may thereby transit to acceleration regimes other than ambipolar expansion. This is expected to affect observables including maximum ion energies and spectral distributions, as observed in simulations and experiments.

As an example, a hydrogen sphere of 500 nm diameter and $n_e = 340 n_c$ would require $a_0 > 32$ (according to Eq. (5.11) and Fig. 5.7), to instantaneously induce a Coulomb explosion without expanding the target during the laser interaction. This corresponds to a few-cycle limit; with a single cycle of 3.3 fs duration and neglecting the target expansion during the laser interaction, the intensity required to heat all target electrons (according to Eq. (5.18)) is even larger, $a_0 > 293$. The explosion could then yield maximum proton kinetic energies of 128 MeV. Figure 5.8a and b show, that already in the single cycle limit, the constraints will realistically be slightly reduced in comparison to the 'ideal' case (Fig. 5.7) by the dynamic target expansion.

If the interaction of the same target is now considered with only $a_0 = 12$ ($k_B T_e = 3.85$ MeV), but *with* the dynamic target expansion during a pulse of 150 fs duration (cf. Fig. 5.8c), the target would expand to a diameter of $\sim 6.3\,\mu$m during the interaction with the laser pulse. This would lead to a Coulomb potential of the fully charged sphere Eq. (5.4) of only 10 MeV. The required field a_0 at that point would only amount to $a_0 = 9$, which in return implies that the Coulomb threshold is factually overcome dynamically during the interaction (as the laser provides $a_0 = 12$). Based on Eq. (5.11) we find that this occurs after ~ 75 fs. From there on, we expect Coulomb repulsion to dominate the ion-motion. At that time, the target has expanded to a momentary diameter of 3.5 μm with a Coulomb potential of the fully ionized sphere of 18.4 MeV. In addition, ions already carry the momentum of

---

[1]In the following calculations and examples, we evaluate and refer to proton sound velocities. Note however, that in comparison the sound velocity for fully charged carbon ions reduces only by a factor $1/\sqrt{2}$ which was found to barely influence these considerations—which are based on estimates anyway.

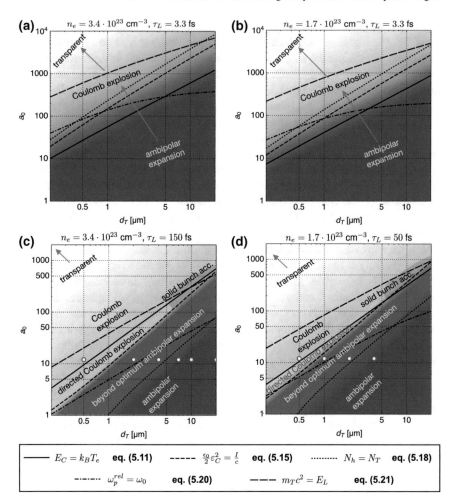

Fig. 5.8 Expansion regimes for temporal flat-top laser-pulse of finite duration with dynamic target expansion. Laser-ion acceleration mechanisms for isolated spheres with specified initial densities and pulse durations considering a temporal flat-top laser-profile with a laser wavelength of 1 μm. Regimes are given as colored areas. Lines separating regimes by initial target-diameter d_T and laser-intensity are explained in the legend. Figure **a** and **b** make the connection to the case without any dynamic expansion in Fig. 5.7 by assuming a very short (3.3 fs) pulse duration. Red dots mark parameters used later in **c** experiments and in **d** simulations respectively. Simplification Eq. (5.19) was not used here

the pre-expansion of ∼3.85 MeV, such that the total final energy for protons in this hybrid acceleration scheme amounts to ∼22.25 MeV.

In a second example we consider the optimized regime, where $N_h = N_T$ and the target does not turn transparent. For the same laser and target density, this coincides with an initial target diameter of 4 μm that expands to 9.8 μm in the interaction and yields 39 MeV in the pre-expanded case (with Eq. (5.8)) instead of 41 MeV in the calculation without pre-expansion. Naturally, the effect of pre-expansion is smaller in larger targets, as the relative pre-expansion is smaller.

Because the target in this second example stays opaque during the interaction (despite $N_h = N_T$), the actual hot electron number may still be smaller than the number of electrons contained in the target and the energy-scales may therefore be effectively smaller in reality. Meanwhile, a target that would turn transparent (such that $N_h = N_T$ could be established), requires more elaborate considerations with respect to electron heating temperature and efficiency. This dilemma makes it generally difficult to predict correct ion kinetic energies for this example in a laser-plasma experiment; it is best resolved via numerical simulation and experiment. Before finally doing so, the influence of a more realistic temporal pulse shape as a flat top shall be considered.

Gaussian Temporal Intensity Distribution

Taking into account the more realistic temporal Gaussian intensity-distribution of the laser pulse in the next step, we use a pulse envelope given by

$$a_0^2(t) = a_{0,p}^2 \cdot \exp\left(\frac{-4 \ln 2 \cdot t^2}{\tau_L^2}\right). \tag{5.22}$$

Note that in this definition the peak intensity corresponds to a normalized field strength of $a_{0,p}$ and occurs at $t = 0$. The parameter τ_L is the full-width at half maximum duration of the intensity-distribution. As our estimates rely on instantaneous comparisons of the laser-field with the electrostatic situation of the target, we integrate the relevant parameters to assess these situations at peak-intensity. This means for example, that in case of the first condition for Coulomb explosion Eq. (5.11) we assess, whether the laser can remove all electrons at the peak intensity, given the instantaneous radius of the pre-expanded target at this moment $r_T^{t=0} = r_T^{t=-\infty} + \int_{-\infty}^{0} c_s(t)dt$. The ion sound velocity $c_s(t) = (Z_i k_B T_e(a_0(t))/m_i)^{0.5}$ is evaluated at the momentary laser intensity. The same instantaneous comparison at peak-intensity is used for the second condition for Coulomb-effects, which is based on the balance of electrostatic and radiation pressures Eq. (5.15). To assess the constraint on optimized ambipolar expansion ($N_h = N_T$, Eq. (5.17)) we integrate the energy accumulated on the expanding target up to the peak-intensity as $E_L = (\eta n_c m_e c^3 \pi/2) \cdot \int_{-\infty}^{0} a_0^2(t) r_T^2(t)dt$. The same energy-integration is used to assess the equality of laser-energy with target-mass to determine regime boundaries of solid-bunch acceleration Eq. (5.21). Finally, for the occurrence of relativistic transparency we evaluate Eq. (5.20) with the target density inferred from the instantaneous radius $r_T^{t=0}$ at the peak intensity. The such updated parameter space is summarized in Fig. 5.9, in fact showing only minor deviations from the treatment using the flat-top temporal laser-profile shown in Fig. 5.8.

Coming back to our previous example of the Coulomb explosion of a 500 nm diameter and $n_e = 340 n_c$ initial solid density target, we now investigate the influence of the temporal Gaussian pulse distribution. We consider a laser pulse of 150

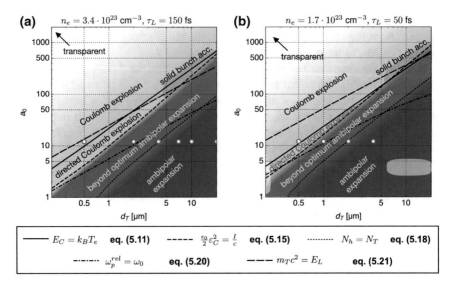

Fig. 5.9 Expansion regimes for temporal Gaussian laser-pulse of finite duration. Laser-ion acceleration mechanisms for isolated spheres with specified densities and pulse durations considering a temporal Gaussian laser intensity profile (Eq. (5.22)) with laser wavelength of 1 μm. Regimes are given as colored areas. Lines separating regimes by initial target-diameter d_T and laser-intensity are explained in the legend. Red dots mark parameters used later in **a** experiments and in **b** simulations respectively. Simplification Eq. (5.19) was not used here. White shaded areas in **b** indicate previous experimental efforts, each individually using a comparably narrow target-size range [21, 35, 36]

fs FWHM duration and peak intensity of $a_0 = 12$ with plane-wave spatial distribution. Equation (5.11) predicts that Coulomb conditions are reached already 40 fs before peak interaction. The Coulomb potential for the fully charged sphere at the momentary target diameter of 4.3 μm is 15 MeV. The momentary ion sound velocity driving the pre-expansion contributes additional 3.4 MeV. This leads to the total energy of 18.4 MeV expected from the scenario, further reducing the predicted energy compared to considerations for the flat-top intensity profile presented before.

In the second example, the 4 μm diameter target, the results of the calculation are basically unaltered from those using a flat-top temporal laser pulse.

Summarizing the laser-driven ion acceleration off micro-plasmas and its dependence on laser-pulse duration, we have discussed three steps. First, the instantaneous discussion neglects any effect (i.e. expansion) that occurs already during the interaction with the laser pulse, and shows an ordered transition between ambipolar expansion and Coulomb explosion. Such treatment is valid only for extremely short laser pulses (few-cycle limit). Second, if the target expansion during the laser pulse is taken into account with a flat-top temporal intensity profile and the ion sound velocity, a much richer set of acceleration mechanisms occurs. Third, this set of acceleration mechanism is maintained, even if a more realistic Gaussian temporal intensity profile is used. This is an important result, as it realistically allows to access acceleration mechanisms other than quasi-neutral and ambipolar at much reduced intensities (compared to the case neglecting dynamic target expansion), which are readily available in laboratories.

A similar treatment as done here for the temporal laser-profile may be relevant for the spatial intensity distribution, if the target exceeds the spatial scale of the laser (e.g. its full width at half maximum) during the interaction. Since this will be relevant only for larger and thus generally less interesting targets, we limit this discussion here to stating the fact.

5.2 Particle-in-Cell Simulations

In modeling the complex reality of intense laser pulses interacting with solid density mass limited targets, numerical particle in cell (PIC, [37, 38]) simulations proof powerful to gain a more detailed theoretical knowledge.

From an experimentalist point of view, the most intriguing feature of such simulations is the full simulation of the laser-plasma interaction and the comparably simple access to extensive and time-resolved diagnostics for interaction products. Particularly in combination with the verification of single points in the analytical parameter-space given in the section above, they enable a good understanding of relevant processes. This chapter discusses simulations that were performed in close context with experiments presented later. All simulations were performed at the chair of Prof. H. Ruhl (LMU) by V. Pauw and K.-U. Bamberg on the SuperMUC computer using the Plasma Simulation Code [38].

A scan of target-size has been performed while keeping other parameters, especially the laser, constant throughout the scan. The simulated setup models the interaction of a linearly polarized laser pulse with a spherical target consisting to equal parts of fully ionized carbon and hydrogen. The electron density is set to $n_e = 1.7 \cdot 10^{23}$ cm^{-3} corresponding to half solid density of polystyrene, modeled with 9155 particles per quasi-particle at 16 nm spatial resolution. The simulation domain contains $2688 \times 1536 \times 1536$ cells and all boundaries are open. The simulated laser pulse has a Gaussian pulse shape with 55 fs FWHM duration and a 8.3 μm FWHM Gaussian transverse intensity profile unless stated otherwise. The laser peak intensity corresponds to $a_0 = 12$ at a laser wavelength of $\lambda = 1$ μm. For this chapter, simulations were performed with target diameters of 500 nm, 1 μm, 2 μm and 4 μm. In order to ensure numerical stability, particles are initiated with a homogeneous thermal plasma temperature of 10 keV. This corresponds to the estimated integrated energy level expected to arrive at the target prior to peak interaction.[2] Simulation parameters have been chosen closest possible to values available in the experimental campaign presented in Sect. 5.3 taking into account computational limitations.

---

[2]Assuming a peak intensity of 10^{20} W/cm^2, ASE level of 10^{-8} and ASE duration of 10 ns, the energy absorbed in a particle with radius 250 nm will range in the 10 μJ level, equivalent to about 10 keV average energy per particle.

This section exclusively presents 3D3V simulations, since preliminary studies using intrinsically faster and cheaper 2D3V simulations were proven inaccurate for modeling the physical reality of a sphere target interacting with a high power laser [39]. One reason for this is the misrepresented target-geometry, leading to unphysical electron heating and Coulomb fields in two dimensions and consequently unreliable results regarding ion and electron kinetic energy distributions. This same geometric misrepresentation is also responsible for a false modeling of expansion effects (during and after the laser-pulse).

5.2.1   Electrons

The projections of electron density in Fig. 5.10 show periodicities of $\lambda/2$ and λ in the planes perpendicular and parallel to the linear laser polarization in y-direction, caused by the $\vec{j} \times \vec{B}$-force. As the laser intensity is well beyond $a_0 = 1$ electrons are mainly scattered in laser-propagation direction, once they are injected into the laser-field. For the 4 μm sphere, electrons are carried away from the target in a train of sub-fs and dense (few-% critical density) bunches. These electron bunches are an interesting subject on their own right; in a complementary experiment at the few-cycle light wave synthesizer LWS20 [40], we were able to demonstrate the sub-cycle control of relativistic electron ejection from a nanotarget via the carrier envelope phase of the laser pulse [41, 42]. Meanwhile, the critical density surface at which plasma frequency and laser frequency are balanced remains *relatively* close to the original target-vacuum interface. More than 85% of electrons remain within the initial sphere-volume. For a 500 nm target the picture looks distinctively different; while dense electron bunches are carried away similar to the large sphere-case, the target disintegrates during the interaction, with less than 10% of the electrons remaining in the initial sphere-volume. The critical density is not maintained to the end of the laser-pulse duration.

Electron kinetic energy spectra for four different target diameters, evaluated at 15 fs after peak interaction, Fig. 5.11, all show a broadband decay. In addition we observe the transition from a steep exponential decay in the 4 μm target, corresponding to the heating in the cold plasma approximation (i.e. with strong Coulomb background), to a much flatter decay towards higher energies for smaller targets, which ultimately dominates for the target with 500 nm diameter. This transition can be identified with the transition of heating mechanisms that was readily implemented in the analytical derivation of regime boundaries earlier (Sect. 5.1.3).

Our analytical modeling relied on the assumption of electrons being heated homogeneously, and characterizable by a single number, their average kinetic energy (temperature). As expected, this assumption is challenged by the observed electron acceleration, with prevalence in laser-propagation direction. We address this issue with a closer look at simulation results. A simple means to quantify the inhomogeneity is the evaluation of the mean kinetic energy of electrons (here sometimes referred to as temperatures) sitting in different spatial zones; for the expansion of the

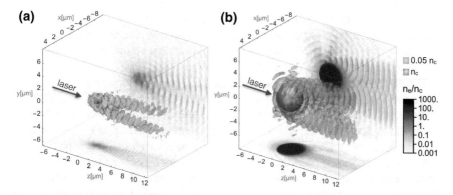

Fig. 5.10 Simulated electron density. Electron density in a 3D3V PIC simulation at 15 fs after peak interaction for a target that is initially **a** 500 nm and **b** 4 μm in diameter. 3D plot gives contours for critical and 5% critical density, 2D plot gives respective projections to the polarization planes using a 160 nm thick cut-slice through the center of the interaction

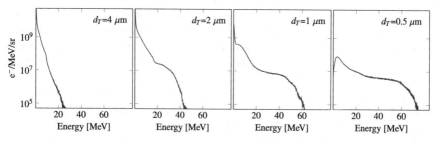

Fig. 5.11 Simulated electron spectra. Electron spectra at 41 fs after the peak-intensity (time chosen such that all electrons are still contained in the simulation-box)

ion population it is most relevant to distinguish electrons sitting inside and outside of this population.

This comparison is summarized in Fig. 5.12. Figure 5.12a–d serve to qualify up to which time the calculation of spectra is meaningful, i.e. not too many electrons have left the simulation box. Figure 5.12e–h show the average kinetic energy of electrons inside of the ion population, outside of the ion population and overall. For the 4 μm target at 15 fs after the peak interaction, the mean kinetic energy of electrons inside the ion population is just 0.08 MeV. Outside of the ion-cloud it amounts to 3.8 MeV, in fact very close to the ponderomotive scaling Eq. (2.45). As most electrons stay within the initial (opaque) target-volume, the number of heated electrons is much smaller than the number of target-electrons. Hence, the overall temperature is dominated by relatively cold electrons which redistribute the energy of hot electrons among themselves leading to smaller overall temperatures.

For the 2 μm target, theory predicts the equivalence $N_h = N_T$, i.e. the hot-electron number equals the number of available electrons. In fact, the temperature inside the ion population is raised close to the MeV level, although the target stays opaque.

With a temperature of 7.6 MeV at 15 fs after the pulse, and up to 12.3 MeV at later times, the population outside of the ion-population reaches several times the ponderomotive scaling (Eq. (2.45), 3.8 MeV), but stays below the ponderomotive potential of a free electron (Eq. (2.32), 18 MeV).

For the 1 μm target, we observe a rise in the average kinetic energy of electrons inside the ion population up to to 2.6 MeV, indicating the beginning transparency of the target. Outside of the ion population, the temperature is further raised to 9.33 MeV at 15 fs after the peak. This outside temperature reaches up to 18 MeV at 95 fs after the peak interaction, corresponding to the free electron scattering, before high energy electrons start to leave the simulation box in significant numbers. The average kinetic energy of all electrons is increased to a maximum of 4.2 MeV. Meanwhile, the electron depletion (Fig. 5.12i–l) from the ion population is still rather ineffective for this target, and consequently the Coulomb repulsion is not expected to be fully effective either. The (mildly) increased overall electron temperature at reduced target size relates to our discussion of the 'beyond optimum ambipolar expansion'.

Eventually, for the 500 nm target we observe heating inside and outside of the ion population to far beyond the ponderomotive scaling for cold plasmas (Eq. (2.45), 3.8 MeV). The average kinetic energy is dominated by the outside population, because almost all electrons are now removed from the target, which does no longer represent a sufficient Coulomb attractor for these electrons to hold them back. As the target eventually turns transparent and electrons can be efficiently accelerated in the co-propagating transmitting laser-field, which is no longer shadowed by the target, they gain energies slightly exceeding the ponderomotive potential of a free electron in the corresponding field strength (Eq. (2.32), 18 MeV). The electron depletion reaches 85% already shortly after the peak interaction, when the target is still relatively small. Correspondingly the Coulomb explosion can be expected to contribute strongly to the ion acceleration, which may be further boosted by the use of a multi-species target.

Concluding this discussion, it is clear that the isotropic distribution of electrons and the homogeneous heating are rough assumptions, used to derive analytical insights to the ion acceleration. Especially for large targets, the assumption would mean a presence of near-critical density (or even solid-density) hot electron populations in the complete target-volume. This contradicts simulated observations, where the target core of large targets stays rather cold, while electrons are heated only close to the target surface. The example well represents the difficulties that quickly arise when trying to estimate hot electron number densities and temperatures, let alone ion parameters, that will arise from such a complicated scenario. Despite these difficulties, the transition from the bulk plasma heating at large targets to the ponderomotive scattering of quasi-free electrons in smaller targets was predicted by the simplified analysis and recognized in the simulations, e.g. via an increased electron temperature. The effect is promoted by the sufficient laser energy ($N_h > N_T$), by induced transparency of the target, and by the reduced Coulomb background.

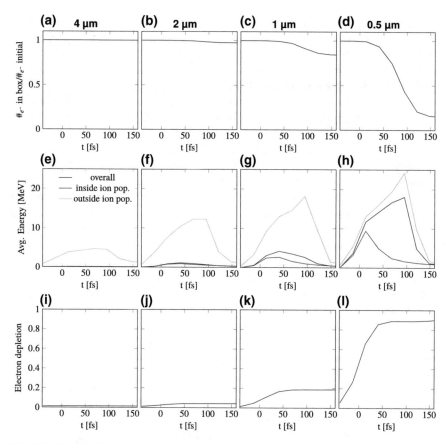

Fig. 5.12 Spatial distribution of electrons and electron energy. a–d Fraction of electrons that are still in the simulation box. **e–h** Average kinetic energy of electrons inside of the ion population, outside of the ion population and combined versus time. **i–l** Fraction of electrons depleted from the ion population of the target

5.2.2 Ions

Following the discussion of electron dynamics, the most exciting question is its impact on the ion-acceleration. Figure 5.13 shows ion spectra at the end of the simulation, 230 fs after the peak interaction with the peak-intensity of the laser. Plotted spectra include all ions within the simulation, i.e. they are not spatially filtered. The first observation is, that proton energy is smaller for the 4 μm as compared to the 2 μm target. This was to be expected due to the small overall electron temperature in the larger target. For the 2 μm target, a maximum kinetic energy of 25 MeV is reached. The maximum proton kinetic energy for the 1 μm target stagnates as compared to the 2 μm target, which again indicates the beyond-optimum ambipolar expansion as expected from Fig. 5.9. The effects of the smaller target and the increased average electron kinetic energy counterbalance each other to yield a very comparable maximum proton kinetic energy in both cases. Note that the expectation from the simple model for the 4 μm target, 38 MeV kinetic energy in protons, is never met. Going further, the maximum proton kinetic energy for the 500 nm target is slightly increased as compared to the 1 μm target, indicating, that indeed the increased electron temperature—rising even faster once the target turns transparent and reaches Coulomb conditions—can counteract the effects of reduced target-size for the maximum energy.

These observations explicitly contrast the behavior of maximum kinetic energy predicted by the simple model Fig. 5.6, which showed monotonically increasing maximum ion kinetic energies towards larger targets, when assuming homogeneous and instantaneous heating. In consequence, models as used in Fig. 5.6 can hardly be used as a predictive tool in terms of maximum ion energies or exact spectral ion distributions for particle-in-cell simulations or for experiments, particularly when scanning through acceleration mechanisms.

For carbon ions, the maximum kinetic energy of 12 MeV/u is reached for the 1 μm target and shows a decreasing trend towards both sides (smaller and larger targets). Due to their larger inertia in comparison to the protons, carbon ions expand to a smaller extent during the laser pulse. In the complex interplay of target size and electron temperature, the increased electron temperature may lead to the slightly increased carbon ion kinetic energy. In the smallest target, the protons are simply more mobile than the carbon ions and hence more efficient in extracting the energy from the Coulomb field.

The second observation is the electrostatic shock between protons and carbon-ions. This was predicted qualitatively in the simple model Fig. 5.6, and gains significance towards smaller targets because the Coulomb-repulsion becomes increasingly effective. In each ion species we observe two parts of the population: that are ambipolar-driven ions evincing the quasi-exponential decay, and shocked populations at the velocity interface, leading to compression of the high-energy carbon and proton populations (forming peaks, as known from bulk foil targets containing multiple ion species [43–45]). In case of the smallest target, where the electron spectra already hint the occurrence of Coulomb explosion, we observe the expected quasi-monoenergetic distribution of protons alongside a carbon-ion spectrum that resembles expectations for a single-species Coulomb explosion (cf. Fig. 5.6c). In this case we can test our example calculation of the hybrid Coulomb explosion as introduced earlier in the theory section. According to this calculation, Coulomb conditions are met 12 fs prior to the peak interaction, where the target has a momentary diameter of 2 μm. The corresponding Coulomb energy is 16.1 MeV which adds to the pre-expansion kinetic energy of 3.6 MeV. The total of 19.6 MeV proton kinetic energy comes close to the observed peak kinetic energy. The estimate contrasts the expectation of ion energies from such target in a pure Coulomb explosion without pre-expansion of 64.4 MeV.

The change of mechanisms is also reflected in the density distribution of ions, already during the interaction. Figure 5.14 shows a snapshot of density distributions at 15 fs after the peak interaction. In case of the 500 nm target, low-energy pro-

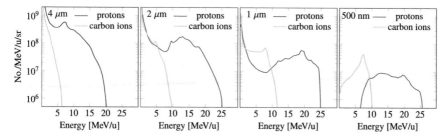

Fig. 5.13 Ion spectra at the end of the simulation, 230 fs after peak interaction for targets of varying diameter

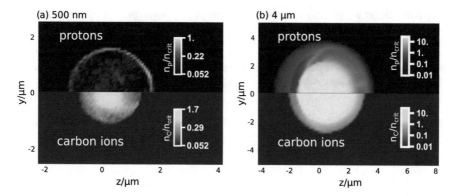

Fig. 5.14 Simulated ion densities. Zoomed snapshots of the PIC simulation at 15 fs after the peak interaction for **a** a 500 nm diameter target and **b** a 4 μm diameter target. The laser pulse is polarized along the z-coordinate with a peak normalized amplitude of $a_0 = 12$ and FWHM spot size of $d_L = 8.3$ μm. The projections show the proton ($y > 0$) and carbon ($y < 0$) density distribution for a 160 nm thick cut-slice through the target center

tons are pushed so violently out of the carbon-occupied volume, that they outrun protons sitting initially further outside and gaining initially more kinetic energy via ambipolar expansion. At the instant of the snapshot, most of the protons form a dense and compact shell around the carbon core. After that process, carbon ions follow a Coulomb explosion like behavior, while acceleration for protons stagnates. For the 4 μm target at the same instant, protons and carbon ions still overlap and expand largely independently from each other. From both plots it is also evident that ion acceleration can still be assumed to be isotropic in both cases, despite the electron-heating itself being partially directional.

Finally note, that observed diameters of ion populations match the predictions based on the ion sound velocity fairly well, which was used as the foundation of the theoretical treatment of dynamic target-expansion (pages 76 ff.), predicting radii of 1.5 μm and 3.3 μm for the proton populations, and 1.1 μm and 2.9 μm for the carbon ions respectively. More generally, simulations indicate that the ion sound velocity as suggested e.g. in [14, 27] indeed serves fairly well to estimate the target

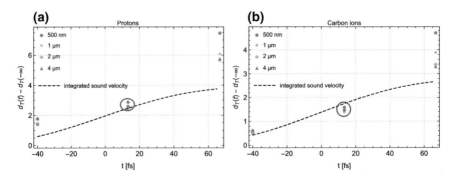

Fig. 5.15 Target expansion versus time. **a** Expansion (i.e., initial diameter subtracted from the momentary diameter) of the proton population (in μm), evaluated at the outermost boundary of the population orthogonal to the laser pulse propagation for four different initial diameters. The dashed line represents the estimate based on ion sound velocity c_s with Gaussian temporal pulse shape and the red circle marks the region of interest close to peak intensity (t = 0). **b** Same evaluation for (fully ionized) carbon ions, with adjusted ion sound velocity

expansion during the laser pulse interaction, and therewith the relevant acceleration mechanisms, in a straight forward way—regardless of the initial sphere diameter and the final acceleration mechanism (cf. Fig. 5.15).

5.3   Experiments at TPW: Target-Size Scan

This section presents experiments aimed to demonstrate the variety of acceleration mechanisms accessed via dynamic target expansion.[3] A 2014 experimental campaign was carried out at the Texas Petawatt laser performing a target-size scan across the laser focal diameter. This section grew out of our related publication Ref. [46].

5.3.1   Setup

Laser pulse and target A top view of the experimental setup is shown in Fig. 5.16. Besides the optical setup for the Paul trap, the ion gun, the illumination laser of the trap system and the trap itself, the experimental vacuum chamber was equipped with the additional elements required to conduct the high power laser-plasma experiment. The laser is transported to the target, using the plasma-mirror setup depicted in Fig. 3.4. Behind the laser focus a microscope (focus diagnostics, FD) was used to

---

[3] My contribution: I initiated the experiment writing the corresponding proposal, planned the experiment and logistics, executed the setup and experiment as principle investigator. I supervised the team of 5 scientists (not counting TPW-staff), and organized the daily shot-plan. I performed the data analysis and interpretation.

optimize the laser-intensity distribution in the focus (i.e. minimizing astigmatism). The microscope produced a 20X magnified image of the laser focus relayed to a camera outside of the chamber. For adjustment, the laser was used at much lower laser energies (mJ level) and higher repetition rates (2.5 Hz) than actual full-power specifications, for practical reasons. The plasma mirror ignites only (and then needs to be changed) for each full-system shot. Before ignition, it still reflects sufficiently well ($R \sim 10^{-3}$) to adjust the focus at moderate intensity. The three-dimensional overlap of the target with the laser-focus was ensured using the same FD microscope; the target-position was brought to a fixed position in the camera image. For the time of focus adjustment, the target was by means of trap-voltages driven out of that position to avoid any laser-induced target-changes before the interaction. The laser-focus was brought to the exact same fixed position. Finally, the target was repositioned to that position. In fact, the depth of field of the microscope and its lateral resolution easily suffice to position the target in the highest intensity region (i.e. within the Rayleigh range of $\sim$20 μm and FWHM spot size of 10 μm in the current experiment).

During the trapping process and directly before system-shots, the focus diagnostics was driven to a save position as depicted in Fig. 5.17, preventing damage via ion sputtering (by the ion gun) and by laser- or particle-beams from the super-intense interaction respectively. For the same reason, automatically triggered mechanical shutters closing just milliseconds before the interaction were introduced to the illumination laser of the optics and to the optical setup of the trap, which was shown not to influence the quality of target positioning.

Ion Diagnostics The main diagnostics for the laser-plasma interaction in this experiment was the measurement of energetic ions. Downstream of the target, a wide angle ion spectrometer, WASP (Appendix B.1) was used to measure 1D-angular resolved differential ion kinetic energy distributions for each laser shot, effectively in a range of ±2.2° around the central axis in a plane perpendicular to the laser polarization. In the detection plane, a Fuji Film image plate detector of type BAS-TR (IP) was used to detect the signal. The IP was covered by 100 μm Aluminum and partly covered by 1 mm thick CR39 nuclear track detectors in order to calibrate particle number and to distinguish carbon ions from protons via penetration depth. The X-ray-dominated background signal plus the detector noise level have been estimated conservatively and subtracted from the spectrum for each shot.

As for most high-power laser-plasma experiments, the greatest challenge in the setup was the strategic accommodation of *all* movable components at the precision of better than 100 μm. This range is the typical field of view for both the focus and optical Paul trap diagnostic, and also the order of the entrance-slit opening of the WASP. Many parts require motorized or manual translation stages, as they need to be adjusted prior to the shots. Figure 5.17 shows a photograph of the crowded experimental setup including the trap in its center.

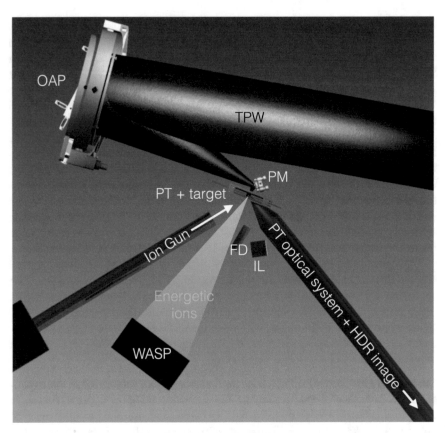

Fig. 5.16 Top view schematic of the experimental setup at TPW. Key components are the Paul trap (PT) itself including its optical system, the illumination laser for the Paul trap (IL), the focus diagnostics microscope (FD, depicted in its safe position), the plasma mirror (PM) and the slit and dipole magnet of the ion wide angle spectrometer (WASP). For reference, the initial beam diameter is 220 mm

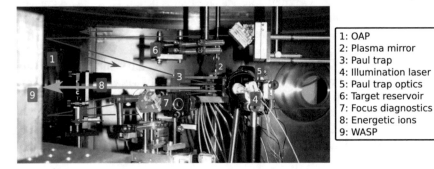

1: OAP
2: Plasma mirror
3: Paul trap
4: Illumination laser
5: Paul trap optics
6: Target reservoir
7: Focus diagnostics
8: Energetic ions
9: WASP

Fig. 5.17 Setup photograph. Photograph of the experimental setup in the TPW vacuum chamber

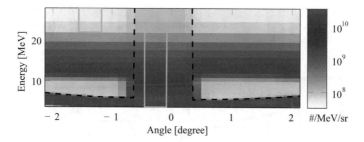

Fig. 5.18 Angularly resolved proton kinetic energy spectrum. Recorded for a 10 μm target, recorded with the WASP. Regions marked by the black-dashed line were shadowed by 1 mm thick CR39 detectors in addition to the 100 μm Aluminum shield. Green regions are one possible set of regions used to extract angle-averaged proton-spectra in Fig. 5.19

5.3.2 Results

Commercial targets [47] with diameters of 520 nm, 1.96 μm, 4 μm, 7 μm, 10 μm and 19 μm were irradiated, spanning more than four orders of magnitude in target mass and electron/ion number.[4] Specified target diameters had deviations of only 2% rms within a sample. The Texas Petawatt laser (TPW) delivered parameters given in Table 3.1. Overall 19 out of 33 shots produced an evaluable signal and were used in the analysis, the major issue for non-used shots being the focus pointing fluctuation. No significant angular dependence of the signal was observed within the measured range (cf. Fig. 5.18).

Figure 5.19 shows differential proton kinetic energy distributions for best shots (i.e. with highest kinetic energies), obtained for different sphere sizes. For large spheres (19–7 μm), the proton kinetic energy distribution $d^2N/dEd\Omega$ decays exponentially with kinetic energy E, as expected from a cold ambipolar plasma expansion/ quasi-neutral expansion. The data in Fig. 5.19b for the 10 μm sphere are well described by the Maxwell distributed spectral shape Eq. (5.7) with the hot electron temperature of 3.8 MeV obtained from the ponderomotive scaling Eq. (2.45) [49] for the peak-intensity ($a_0 = 12$). This analytical spectrum is represented by the red line in Fig. 5.19b. As expected, only the high-energy part of the measured ambipolar spectrum shows some deviation, including a distinct maximum cutoff energy.

For the large 19 μm spheres, both maximum proton kinetic energy and particle number are significantly reduced. This can be understood intuitively because the sphere significantly exceeds the FWHM diameter of the focus and can thus be regarded as a large cooling volume which dissipates energy and reduces the energy density. For slightly sub-focus sized 7 μm targets we find maximum proton kinetic energies of 28 MeV. For the 4 μm targets we observe comparable energies and start to

---

[4]This subsection is reproduced with minor variations and with permission from the original peer-reviewed article: T. M. Ostermayr et al., Physical Review E, 94(3):033208, (2016). The article is published by the American Physical Society and licensed under a Creative Commons Attribution 3.0 International License.

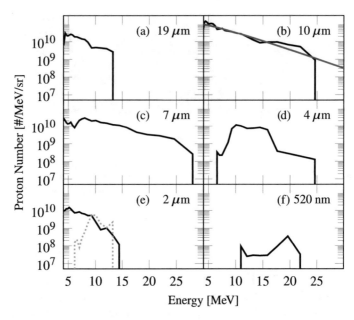

Fig. 5.19 Target size-scan: spatially averaged experimental proton kinetic energy spectra recorded with the WASP. a–f Black solid lines are the proton spectra of best shots (max. kinetic energy) for the respective target size. Gray dotted line shows the second-best shot at 2 μm. Noise and background were evaluated and subtracted for each shot individually. The red line in **b** is the quasi-neutral spectrum for the 10 μm target predicted by [5, 48] at characteristic energy of 3.8 MeV [49]. Figure adapted with permission from original publication [46]

observe non-monotonic distributions for some (including the best) shots. According to earlier arguments, this range represents the optimized ambipolar expansion. The maximum kinetic energies are—similar to the simulation—short of the analytical expectations from Sect. 5.1.3 yielding 48 and 39 MeV for the 7 μm and 4 μm targets respectively. Reducing the target diameter further to 2 μm, the proton kinetic energies decrease significantly, despite non-monotonic behavior occurring for some shots. Here, instead of just stagnating as in the simulations, the decreased target size even leads to a decreased proton kinetic energy in the 'beyond optimum ambipolar expansion'.

For the smallest target used in this experiment, 520 nm, protons are exclusively accelerated to kinetic energies between 11 and 22 MeV while the low-energy detection limit of the spectrometer is at 3.3 MeV. The spectral situation in case of the 520 nm target allowed the distinctive measurement of maximum carbon kinetic energy, which is in the range of 104–121 MeV. This means that the kinetic energy per unit-charge is similar for fully charged carbon-ions and for protons. Overall, we experimentally recovered all key features from the simulation: the quasi-monoenergetic feature in protons, the increase in maximum proton kinetic energy (in comparison to the next larger target), and the maximum carbon kinetic energy.

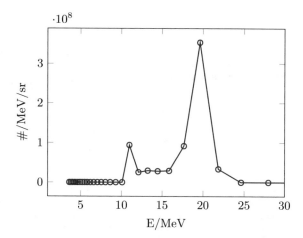

Fig. 5.20 Proton spectrum for the 520 nm target shown in Fig. 5.19f, reproduced with a linear scale

Earlier in this chapter we showed that the Coulomb-repulsion of carbon-ions pushes initially slow protons out of the carbon-rich volume, eliminating the low-kinetic-energy part of the proton spectrum. The such created distinct mono-energetic peak is more clearly highlighted when shown in a linear scale given in Fig. 5.20. The FWHM bandwidth of this spectrum is in fact too small for our spectrometer to be resolved, and can only be given as upper limit of <4 MeV. The conducted experiment with this smallest target coincides with our example discussed for hybrid Coulomb explosion, predicting 18.4 MeV of proton kinetic energy. This matches our experimental finding reasonably well, again contrasting the 128 MeV expected for the pure Coulomb explosion, which would be expected from instantaneous heating/energy deposition. However, and as stressed earlier, such pure Coulomb explosion would require a normalized field of at least $a_0 > 32$ with an instant rise time.

Particle Numbers and Efficiency

We further quantify the numerical efficiency and the efficiency of laser energy conversion to proton kinetic energy. Integration of the proton kinetic energy distributions measured in forward direction (shown in Fig. 5.19) over energies beyond 3.3 MeV (lower WASP detection limit) and multiplying the result with π sr, a conservative estimate for the solid angle of emission extracted from simulations, yields the number of protons emitted from the sphere. The number of protons in the spectrum normalized to the number of protons contained in the target is depicted in Fig. 5.21a, where solid points mark the best shots in Fig. 5.19. Interestingly, for targets with diameters smaller than 4 μm this number approaches the number of protons initially contained in the target, even for this conservative calculation. To calculate the energy-conversion, we divide the kinetic energy contained in registered protons by the incident laser energy. This conversion efficiency η_{lp} is depicted in Fig. 5.21b, reaching values of 4% for a target diameter of 10 μm, matching the focal spot size. We attribute the reduced laser to proton kinetic energy conversion efficiency for decreasing target diameters mainly to geometrically reduced cross sectional overlap with the laser pulse. This is

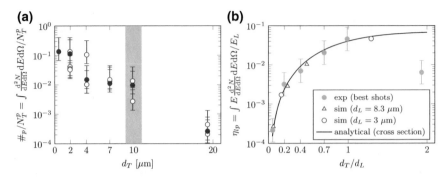

Fig. 5.21 Accelerated target fraction of protons, and laser to proton energy efficiency. a Number of accelerated protons $\#_p$ in relation to the number of protons contained in a sphere N_T^p of the respective size. Solid points mark the respective best shots shown in Fig. 5.19. The blue shaded area marks the laser-focus spot diameter. **b** Laser to proton kinetic energy conversion efficiency η_{lp} plotted against target diameter d_T normalized to the laser FWHM focal spot diameter d_L. Green points are the experimental laser to proton kinetic energy conversion efficiencies for energetically best shots (Fig. 5.19). Blue triangles and circles are results of 3D3V particle in cell simulations with 8.3 and 3 μm FWHM focus diameters respectively. The black curve represents the analytical scaling taking into account the cross sectional overlap of the target with the laser. Figure adapted with permission from [46]

quantified assuming a Gaussian laser spot. The conversion efficiency as a function of sphere diameter is then given as $\eta_{lp}(d_T) = \eta_0 \cdot \Phi(d_T)/\Phi(10 \, \mu m)$, where $\Phi(d_T)$ is the laser energy incident on the sphere with diameter d_T and $\eta_0 = 4\%$ as observed experimentally. For the best shots, the experimental conversion efficiencies (green points) agree well with this simple scaling. For spheres exceeding the focal spot size, the scaling predicts saturating efficiencies comparable to the TNSA case for foils. The slightly reduced values for the 19.3 μm target must be regarded with caution, given our strongly simplified assumption of constant emission angle π sr. Overall the curve shows similarities to predictions by [18] and represents the first experimental confirmation of such scenario. In the range covered by our simulations the energy conversion efficiencies observed in the simulations agree with experiments and with the simple analytic scaling. Simulations performed with a smaller 3 μm FWHM focal spot size indicate that indeed the reduced cross sectional overlap with the laser rather than just the reduced surface of the target is accountable for the reduced efficiency. Hence, highly efficient proton acceleration may be feasible with laser focus and target matched at smaller sizes, relaxing the costly requirement of laser energy.

Earlier considerations, observations and discussions in simulations and experiments were mostly based on the target expansion during the laser-pulse interaction. Therefore, the accuracy of a scaling-law through all simulations and experiments, that fully neglects these effects altogether, is surprising. Within these earlier considerations, the electron heating (in terms of electron number and temperature) was found to be critical for the plasma-expansion, and yet particularly difficult to model for a set of given parameters. The understanding of the simple analytical scaling for conversion efficiency may help to refine models and to resort some of these generally known issues (after all, the electron heating is also crucial in the modeling of plane-foil target interactions) in future work.

For example, one could speculate about the following scenario, based on the hypothesis that the conversion efficiency from laser to proton energy is an image of the conversion efficiency to electrons. With a target of initial radius r_T and electron number density $n_{e,0}$ that pre-expands according to the considerations, the volume accessible to the laser interaction at a given time is $V_{int} = 2\pi r_T^2(t) l_s(t)$, where l_s is the laser skin-depth with proportionality $l_s(t) \propto n_e^{-0.5}(t) \propto r_T^{1.5}(t)$. The number of electrons interacting directly with the laser is $N_{int}(t) = V_{int}(t) n_e(t) \propto r_T^{0.5}$. This number scales weakly with the momentary target radius, and is hence barely susceptible to the target's dynamic change. If—according to this weak scaling and without effective hole boring effects [50, 51] (i.e. the laser pushing the critical surface further into the target, thereby increasing the effective volume and number of accessible electrons)—the same set of electrons was to interact with the laser throughout the entire interaction, and if the maximum energy gained by a single electron was limited by the laser-intensity as found in scalings and simulations, then the energy transferred to the plasma would be (roughly) limited by the initial target conditions as indicated by the analytical scaling of laser to proton conversion efficiency. Such a model divides the target electrons in two populations, one being heated (N_{int}) and the second one remaining cold, consistent with two-temperature models in literature [52]. The limited energy transfer between both electron populations (limited thermalization) is consistent with electron spectra observed in simulations. This consideration challenges the estimate Eq. (2.46) for the hot electron number, which currently represents a standard used in many analytical models for bulk foil targets [53, 54]. In these bulk foil targets however, the hole boring process in the quasi-1D scenario may efficiently and continuously feed new relatively cold electrons to the laser-heater to achieve the final efficiency, which appears to be an important difference to spherical targets.

How does this interpretation fit with previous discussions herein? With large targets the energy estimate Eq. (5.8) was found to overestimate the proton kinetic energy in case of a 4 μm target (where $N_h = N_T$) for our experimental conditions at 39 MeV (page 79) instead of the experimentally observed 28 MeV. Using $N_{int} \approx 1\% N_T$ in place of N_h from Eq. (2.46) to assess the hot-electron density in Eq. (5.8) yields just 21 MeV. Similarly, for the simulated scenario of a 2 μm target (showing up to 25 MeV proton kinetic energy) the prediction is 18 MeV with the new model instead of 32 MeV with the old model. The exaggerated hot electron number in the modeling of thick bulk targets may thus be tackled with the new approach, although it appears in its current form to underestimate the ion kinetic energy. The equivalence of heated electron number and target electron number ($N_{int} = N_T$) now coincides roughly with the emergence of (relativistic) transparency in Fig. 5.9c and d. Importantly, the other estimates on regime boundaries remain intact.

Current work goes towards the refinement and solidification of the novel scaling idea. This may need to implement further modeling of effects like the removal of high-energy electrons, the increasing electron temperature towards smaller targets, and the hole-boring. The use of isolated targets could provide a unique basis to study and iterate such a model via spectroscopic measurements of all the contained particle species. The challenge to fully accomplish this goal is the development of a suitable set of high-resolution and high-dynamic range diagnostics spanning as much of the solid angle as possible. The current version of the modeling is limited by the underlying assumptions.

5.4   Experiments at PHELIX: Directional Micro-plasma Acceleration of Dense Ion Bunches

In the previous section, we used high-contrast short (170 fs) laser pulses to investigate ion acceleration off isolated targets. An exciting question is the effect of using longer laser pulses, that correspondingly drive further target expansion prior to the peak interaction. At the PHELIX laser (GSI Darmstadt), we performed the experiment with 1 μm diameter polymer targets, with about three times longer laser-pulses and with 150 J pulse energy.[5] Figure 5.22 shows the theoretical framework applied to these laser and target conditions. Requirements for reaching highly directed acceleration modes seem much relaxed and even practically accessible. This is true with the initial target parameters (diameter 1 μm, electron number density $3.4 \cdot 10^{23}$ cm^{-3}, $\eta = 0.5$) as in Fig. 5.22a.

No plasma mirror was used in this experiment, and the target pre-expansion induced by the light arriving prior to the main pulse (Sect. 3.2) is expected to be significant. In Fig. 5.22b it is motivated, that even with a pre-expansion (on the nanosecond timescale) of the initial 1 μm target to a diameter of 8 μm (leading to a near critical plasma density), the solid bunch acceleration seems accessible. That is valid even with the reduced coupling efficiency of laser-light to electrons $\eta = 0.1$, expected due to the onset of (non-relativistic) transparency. The true target pre-expansion and the laser-absorption (as a function of time)—and thus the dynamics of the acceleration—are too complex to be fully captured by our model. Much more detailed considerations of these specific dynamics will be given in the thesis of Peter Hilz.

In this experiment, the sphere target was positioned one Rayleigh length outside of the laser focus to reduce the influence of laser pointing stability; the laser intensity in consideration Fig. 5.22 is matched to this setup. Figure 5.23 shows experimental results for nine consecutive laser shots, measured with a WASP spectrometer. Quasi-

---

[5]The experiment leader of this campaign was Peter Hilz, to whom I served as deputy. As such, I participated in experiment planning, setup and conducting the experiment. All data was evaluated and interpreted by Peter Hilz, and represents the core of his PhD thesis. Here, this data is shown, as it completes the picture obtained from the use of isolated targets for ion acceleration. Figures 5.23 and 5.24 are reproduced with permission from the original peer-reviewed article: P. Hilz et al., "Isolated proton bunch acceleration by a petawatt laser pulse", Nature Communications, 9:423, (2018). The article is published by Nature Communications (Springer Nature) and licensed under a Creative Commons Attribution 4.0 International License.

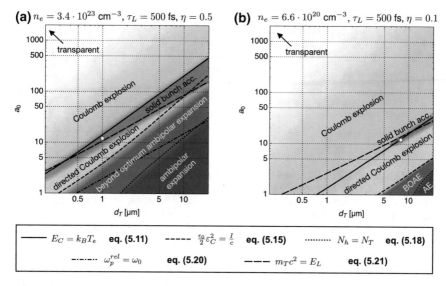

Fig. 5.22 **Acceleration mechanisms with PHELIX parameters**. Prediction on the dominant acceleration mechanism for given laser and target parameters in the PHELIX experiment (red dot). **a** Without expansion prior to the (Gaussian) main pulse. **b** With 8-fold pre-expansion already before the arrival of the Gaussian main pulse, and with consequently reduced absorption efficiency of $\eta = 0.1$

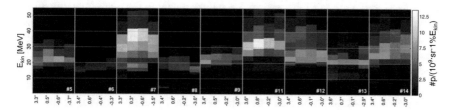

Fig. 5.23 **Measured proton kinetic energy distributions**. Differential proton spectra for consecutive laser shots, for various angles around laser propagation direction (0°). The 'missing' shot 10 was taken as a deliberate 'empty' shot (i.e. without target) for calibration of diagnostics. Figure and caption by courtesy of Peter Hilz, adapted with permission from [55]

monoenergetic proton spectra with peak energies varying in the range of 20–40 MeV, and spectral bandwidths as small as 25% were recorded, thereby showing the robustness of the acceleration mechanism against laser pointing and other fluctuations.

Similar to considerations done for the first presented trap experiment, we can now extrapolate the number of registered particles to the full solid angle (Fig. 5.24a). In contrast to our earlier considerations Fig. 5.21b, here we find a clear contradiction: the amount of particles measured in the spectrum could impossibly be emitted isotropically, simply because the particle number initially contained in the target does not suffice for that by factors of 20–50. This, together with a lack of measured proton signal on CR39 positioned at orthogonal direction with respect to the

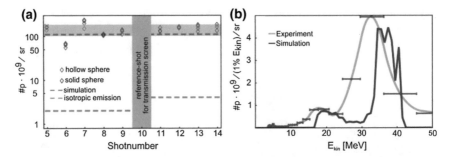

Fig. 5.24 Numerical efficiency and spectrum lineout of experiment and simulation. a Proton number per solid angle from experiment compared to isotropic emission into 4π sr and numerical simulation, the green area represents the standard deviation of the experimental data. **b** Comparison of the double-differential proton kinetic energy spectrum ($\mathrm{d}N/\mathrm{d}E\mathrm{d}\Omega$) for a particle-in-cell simulation and the experiment (shot 11 evaluated at 0.8°). The red error bars indicate the spectrometer resolution. Figure and caption by courtesy of Peter Hilz, adapted with permission from [55]

laser propagation, indicates a large directionality of the accelerated proton 'bunch'. Figure 5.24b further supports this interpretation, by comparing a PIC simulation[6] with a selected measurement, showing good resemblance in spectral shape, energies and particle numbers, corresponding to a highly directed acceleration. The heuristic explanation for this directional mechanism combines two important components: the prolonged interaction with the laser-pulse before reaching the peak-intensity allows to expand and polarize the sphere-target to a greater level than with a shorter laser-pulse, especially as the hole boring velocity $v_h = (\epsilon_0 \mathcal{E}^2 / m_i n_i)^{0.5}$ [50, 51] integrated over the pulse duration τ_L approaches the target diameter. In other words, with the longer laser pulse, the system transits through the directional acceleration regime for a longer time than with a shorter laser-pulse. Thereby protons are drawn to the rear side of the target, while carbons are more inert and lag behind. In addition, the spherical target of near-critical density acts as a phase-object to the laser pulse, leading to a ponderomotive potential that effectively confines electrons close to the laser's propagation axis behind the target and thereby increases the directionality of the entire process (suppressing the isotropic heating). At some point approaching the peak intensity, further effects can become increasingly important; that can be the radiation pressure, or the Coulomb repulsion due to electron removal. In both cases proton acceleration is assisted by the space-charge of carbon-ions; in the pre-polarized setting the acceleration will occur predominantly in the laser propagation direction.

---

[6]This PIC simulation was performed and analyzed by A. Hübl, M. Bussmann, T. Kluge and U. Schramm from HZDR Dresden in close collaboration with P. Hilz, and is part of [55].

5.5 Conclusions

In this chapter, we have introduced the theoretical characteristic ion-energy distributions for different regimes of acceleration, showing distinct behavior. With homogeneous multi-ion-species targets, quasi-monoenergetic spectra can emerge due to the interaction between different ion species, where the ion species with larger charge-to-mass ratio behaves as accelerated test-particles, while more inert species behaves quasi-unperturbed from the single-ion-species case. This sets the basis to use spectral distributions as a probe for the actual accelerating mechanism.

The dependencies on laser- and target parameters were thoroughly investigated in three steps of increasing complexity. If we neglect any expansion during the interaction, a transition from Coulomb explosion to ambipolar expansion is predicted as often discussed in literature [14]. In the next step we added a finite pulse duration considering a flat-top temporal profile. This hybrid model considers the plasma-expansion during the interaction via the ion sound velocity. Already this simple consideration reveals a much richer set of possible mechanisms, readily accessible for our experiments and enabled solely by virtue of the intra-pulse-duration expansion. The system can dynamically transit towards other mechanisms just during the interaction. In a final consideration a more realistic Gaussian temporal intensity profile was considered, showing little difference to the flat-top calculation and thus reconfirming the scenario. This is the first consideration of its kind for isolated targets. The development was triggered by the fact that target expansion during the laser interaction, with current laser intensities beyond 10^{20} W cm^{-2}, significantly alters target parameters (diameter and density). This requires a novel view on data interpretation and contrasts earlier work performed with targets of similar size (in narrower ranges of target diameters (e.g. [21, 35, 36]).

In order to benchmark assumptions regarding electron heating and intra-pulse expansion, simulations were performed, scanning points in the parameter-range sketched in Fig. 5.9. With the current set of simulations, the existence of predicted dynamic acceleration mechanisms, was revealed in signatures in electron- and ion-spectral distributions.

Very similarly looking ion spectra have been found in corresponding experiments at TPW for the expected target-diameters; especially our experiments reproduce maximum kinetic energy behavior and spectral distributions observed in simulations and expected from analysis. This includes the quasi-monoenergetic peak observed for the smallest targets. The evidence establishes a quantitative idea of the laser plasma interaction with a completely isolated mass limited target and a finite pulse-duration. Finally, the newly gained knowledge was leveraged to intentionally and dynamically access a directed acceleration regime in experiments at PHELIX, where indeed the highly directional proton acceleration occurred. The measurement agrees well with simulations showing the predicted directionality and the quasi-monoenergetic spectrum with correspondingly high proton fluence. A complete overview of the quasi-monoenergetic sources measured in this thesis (i.e. within this and the next chapter) will be given in Chap. 7.

This work constitutes the first full theoretical and experimental framework confirming the dependence of laser driven ion acceleration from isolated targets on the target-size and the target-dynamics during the laser-plasma interaction. The basic idea of the theoretical framework may well serve as a reference for future simulations and experimental campaigns aiming to pick one single (or a more limited) set of target and laser parameters to focus on the understanding and optimization of one particular acceleration mechanism. The tunable spectral and spatial distribution of a laser-driven ion beam that is achieved while maintaining other key features of laser-driven ion accelerators (e.g. sub-ns duration and sub-μm-mrad transverse emittance with *tunable* beam divergence) can power several applications. In certain cases, all target-protons were accelerated from the known target. A such defined proton beam is an interesting candidate for special implementations of high resolution quantitative ultrafast proton radiographies. In the difficile field of laser absorption measurements at high intensities, the particle spectroscopy of all particles contained in a target could complement the established methods of collecting photons from the interaction [56, 57]). Another field with potential applications for this kind of plasma source could be laboratory-scale models for astrophysical processes [58]. The micro-plasma accelerator could even serve as the next-generation photo-injector with large and defined numbers of particles and small longitudinal and transverse emittances for next generation (laser-driven or conventionally driven) large-scale accelerator facilities.

Our results are of general relevance for laser driven ion accelerators. The unraveled dynamic acceleration-regimes are paralleled for planar-targets, which are much more commonly used in laser-plasma accelerators. This profound insight stems directly from the derivation of regime boundaries, which are closely related to respective plane-target calculations. To date it is not common practice to account for three-dimensional effects when considering foil targets, even though a tightly focused laser requires exactly that. Importantly, this predicts a realm in between the ambipolar expansion and the Coulomb explosion, with a very complex behavior in terms of ion kinetic energy performance, and growing in parameter-space towards longer pulse-durations. As experiments with foil-targets generally result in much more complex and ambiguous situations, e.g. because of their transverse size exceeding far beyond the focal spot-size and electron currents neutralizing the target, they typically produce complex and ambiguous signals. Meanwhile, recent efforts have identified the dynamic target expansion as one central issue for laser-ion-acceleration from thin plane foil targets as well [59, 60].

References

1. Ditmire T et al (1996) Interaction of intense laser pulses with atomic clusters. Phys Rev A 53:3379–3402
2. Ditmire T et al (1997) High-energy ions produced in explosions of superheated atomic clusters. Nature 386(6620):54–56

3. Göde S et al (2017) Relativistic electron streaming instabilities modulate proton beams accelerated in laser-plasma interactions. Phys Rev Lett 118:194801
4. Weibel ES (1959) Spontaneously growing transverse waves in a plasma due to an anisotropic velocity distribution. Phys Rev Lett 2:83–84
5. Murakami M et al (2005) Ion energy spectrum of expanding laser-plasma with limited mass. Phys Plasmas 12(6):062706
6. Kovalev VF, Bychenkov VY (2003) Analytic solutions to the Vlasov equations for expanding plasmas. Phys Rev Lett 90(18)
7. Dorozhkina DS, Semenov VE (1998) Exact solution of Vlasov equations for quasineutral expansion of plasma bunch into vacuum. Phys Rev Lett 81(13):2691
8. Popov KI et al (2009) Vacuum electron acceleration by tightly focused laser pulses with nanoscale targets. Phys Plasmas 16(5):053106
9. Popov K (2009) Laser based acceleration of charged particles. PhD thesis. University of Alberta
10. Peano F, Fonseca RA, Silva LO (2005) Dynamics and control of shock shells in the coulomb explosion of very large deuterium clusters. Phys Rev Lett 94:033401
11. Kaplan AE, Dubetsky BY, Shkolnikov PL (2003) Shock shells in coulomb explosions of nanoclusters. Phys Rev Lett 91:14
12. Sylla F (2011) Ion acceleration from laser-plasma interaction in underdense to near-critical regime: wakefield effects and associated plasma structures. PhD thesis. Ecole Polytechnique X
13. Peano F et al (2007) Ergodic model for the expansion of spherical nanoplasmas. Phys Rev E 75(6)
14. Murakami M, Basko MM (2006) Self-similar expansion of finite-size non-quasi-neutral plasmas into vacuum: Relation to the problem of ion acceleration. Phys Plasmas 13:012105
15. Sakabe S (2004) et al Generation of high-energy protons from the Coulomb explosion of hydrogen clusters by intense femtosecond laser pulses. Phys Rev A 69(2)
16. Esirkepov T et al (2004) Highly efficient relativistic-ion generation in the laser-piston regime. Phys Rev Lett 92(17):175003
17. Henig A (2010) Advanced approaches to high intensity laser-driven ion acceleration. PhD thesis. Ludwig-Maximilians-Universität, München
18. Limpouch J et al (2008) Enhanced laser ion acceleration from mass-limited targets. Laser Part Beams 26(02):225–234
19. Yu Bychenkov V, Kovalev VF (2005) Coulomb explosion in a cluster plasma. Plasma Phys Rep 31(2):178–183
20. Wei Yu et al (2005) Direct acceleration of solid-density plasma bunch by ultraintense laser. Phys Rev E 72(4):046401
21. Ter-Avetisyan S et al (2012) Generation of a quasi-monoergetic proton beam from laser-irradiated sub-micron droplets. Phys Plasmas 19(7):073112
22. Kluge T et al (2010) Enhanced laser ion acceleration from mass-limited foils. Phys Plasmas 17(12):123103
23. Henig A et al (2009) Enhanced laser-driven ion acceleration in the relativistic transparency regime. Phys Rev Lett 103(4):045002
24. Yin L et al (2011) Break-out afterburner ion acceleration in the longer laser pulse length regime. Phys Plasmas 18:063103
25. Jung D (2012) Ion acceleration from relativistic laser nano-target interaction. PhD thesis. Ludwig-Maximilians-Universität, München
26. Kiefer D (2012) Relativistic electron mirrors from high intensity laser nanofoil interactions. PhD thesis
27. Henig A et al (2009) laser-driven shock acceleration of ion beams from spherical mass-limited targets. Phys Rev Lett 102(9):095002
28. Sokollik T et al Directional laser-driven ion acceleration from microspheres. Phys Rev Lett 103:135003
29. Marx G (1966) Interstellar vehicle propelled by terrestrial laser beam. Nature 211:22
30. Redding JL (1967) Interstellar vehicle propelled by terrestrial laser beam. Nature 213:588

31. Simmons JFL, McInnes CR (1993) Was Marx right? Or how efficient are laser driven interstellar spacecraft? Am J Phys 61(3):205
32. Breakthrough Initiative Starshot. https://breakthroughinitiatives.org/Research/3. Accessed 14 Nov 2017
33. Murakami M, Tanaka M (2008) Nanocluster explosions and quasimonoenergetic spectra by homogeneously distributed impurity ions. Phys Plasmas 15:082702
34. Popov KI et al (2010) A detailed study of collisionless explosion of single- and two-ion-species spherical nanoplasmas. Phys Plasmas 17:083110
35. Ter-Avetisyan S et al (2006) Quasimonoenergetic deuteron bursts produced by ultraintense laser pulses. Phys Rev Lett 96(14):145006
36. Sokollik T et al (2010) Laser-driven ion acceleration using isolated mass-limited spheres. New J Phys 12(11):113013
37. Tskhakaya D et al (2007) The particle-in-cell method. Contrib Plasma Phys 47(8–9):563–594
38. Ruhl H et al (2016) Plasma simulation code. http://www.plasma-simulation-code.net, http://fishercat.sr.unh.edu/psc/index.html. Accessed 25 Feb 2016
39. Pauw V et al (2016) Particle-in-cell simulation of laser irradiated two component microspheres in 2 and 3 dimensions. In: 2nd European advanced accelerator concepts workshop—fEAACg 2015 nuclear instruments and methods in physics research section a: accelerators, spectrometers, detectors and associated equipment, vol 829, pp 372–375
40. Rivas DE et al (2017) Next generation driver for attosecond and laser-plasma physics. Sci Rep 7(1)
41. Cardenas DE (2017) PhD thesis. Ludwig-Maximilians-Universität München
42. Cardenas DE et al (2017) Relativistic nanophotonics in the sub-cycle regime. In: Submitted manuscript
43. Kar S et al (2012) Ion acceleration in multispecies targets driven by intense laser radiation pressure. Phys Rev Lett 109:18
44. Dover NP et al (2016) Buffered high charge spectrally-peaked proton beams in the relativistic-transparency regime. New J Phys 18(1):013038
45. Steinke S et al (2013) Stable laser-ion acceleration in the light sail regime. Phys Rev Special Topics Accel Beams 16(1):011303
46. Ostermayr TM et al (2016) Proton acceleration by irradiation of isolated spheres with an intense laser pulse. Phys Rev E 94(3):033208
47. Microparticles GmbH, Berlin. http://www.microparticles-shop.de. Revisited 30 Jun 2017
48. Kovalev VF, Yu Bychenkov V, Tikhonchuk VT (2002) Particle dynamics during adiabatic expansion of a plasma bunch. J Exper Theor Phys 95(2):226–241
49. Wilks SC et al (1992) Absorption of ultra-intense laser pulses. Phys Rev Lett 69:1383–1386
50. Robinson APL et al (2009) Relativistically correct hole-boring and ion acceleration by circularly polarized laser pulses. Plasma Phys Controll Fusion 51(2):024004
51. Schlegel T et al (2009) Relativistic laser piston model: ponderomotive ion acceleration in dense plasmas using ultraintense laser pulses. Phys Plasmas 16(8):083103
52. Kaluza MC (2004) Characterisation of laser-accelerated proton beams. PhD thesis. Max-Planck-Institut für Quantenoptik and TU München
53. Daido H, Nishiuchi M, Pirozhkov AS (2012) Review of laser-driven ion sources and their applications. Rep Progress Phys 75(5):056401
54. Fuchs J et al (2005) Laser-driven proton scaling laws and new paths towards energy increase. Nat Phys 2(1):48–54
55. Hilz P et al (2018) Isolated proton bunch acceleration by a petawatt laser pulse. Nat Commun 9(1):423
56. Ping Y et al (2008) Absorption of short laser pulses on solid targets in the ultrarelativistic regime. Phys Rev Lett 100(8):085004
57. Bin JH et al (2017) Dynamics of laser-driven proton acceleration exhibited by measured laser absorptivity and reflectivity. Sci Rep 7:43548
58. Albertazzi B et al (2014) Laboratory formation of a scaled protostellar jet by coaligned poloidal magnetic field. Science 346(6207):325–328

59. Bagnoud V et al (2017) Studying the dynamics of relativistic laser-plasma interaction on thin foils by means of Fourier-transform spectral interferometry. Phys Rev Lett 118:255003
60. Bulanov SS et al (2016) Radiation pressure acceleration: the factors limiting maximum attainable ion energy. Phys Plasmas 23(5):056703

Chapter 6
A Laser-Driven Micro-source for Simultaneous Bi-modal Radiographic Imaging

After the investigations presented in the previous chapter, a natural step was to use micro-targets in imaging applications, exploiting their small source size, their large divergence and their unique mixed radiation field composed of ions and X-rays that are geometrically separated from laser and electron beams. The presented experiments were performed in 2016 at the Texas Petawatt laser using commercial tungsten needles [1] as targets.[1]

6.1 Source Characterization

Before applying any particle beam to radiographic imaging, it is useful to explore its spectral and spatial characteristics. A sketch of the experimental setup is depicted in Fig. 6.1. The targets were positioned upright in the 5 μm FWHM laser focus using the focus diagnostics microscope explained earlier. The focus spot was positioned such that the local needle-diameter matched the focal spot diameter, in order to optimize conversion efficiency of laser energy to X-ray and ion beams. The linear polarization was aligned parallel to the needle.

The emission geometry of protons is expected predominantly normal to the target surface, i.e. in the horizontal plane, with an opening angle along the vertical dimension similar to emission angles from foil targets (i.e. around 10° [2]).

X-rays are expected to be emitted in the full solid angle. Diagnostics for photons and protons were positioned in the horizontal plane around the target, and varied for the specific tests of spectral and spatial characteristics.

---

[1]My contribution: I initiated the experiment and wrote the proposal. I planned the experiment and logistics. I executed the setup and experiment as principle investigator, supervising a team of 5 scientists (not counting TPW-staff), and organized the daily shot-plan. I also performed the data analysis and interpretation unless where stated explicitly otherwise.

© Springer Nature Switzerland AG 2019
T. Ostermayr, *Relativistically Intense Laser–Microplasma Interactions*,
Springer Theses, https://doi.org/10.1007/978-3-030-22208-6_6

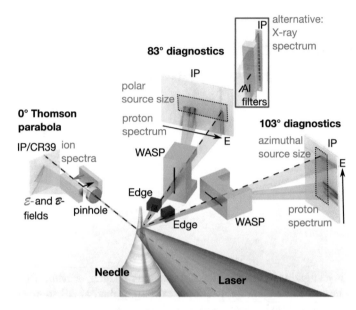

Fig. 6.1 Setup for source characterization. Schematic of the experimental setup for spectral and spatial characterization of particle beams. Ions are indicated by green color, X-rays are indicated by blue color. The diagnostics are two Wide angle spectrometers (WASP) and a Thomson parabola spectrometer using imaging plates (IP) and CR39 nuclear track detectors

A Thomson parabola spectrometer measured ion kinetic energy distributions in laser propagation direction. It consists of a pinhole (300 μm diameter) followed by $\mathcal{B}$-field ($\mathcal{B} = 450$ mT, $L_{\mathcal{B}} = 100$ mm, Drift $= 300$ mm) and parallel $\mathcal{E}$-field ($\mathcal{E} = 20$ kV/3 cm, $L_{\mathcal{E}} = 420$ mm, Drift $= 60$ mm). As detectors for this diagnostics we used CR39 and imaging plates (cf. Appendix B.1).

In the horizontal plane, at 83° with respect to the laser propagation, a combined point-projection edge-image and WASP detector ($\mathcal{B} = 0.1$ T, $L_{\mathcal{B}} = 100$ mm, Drift $= 500$ mm) was used to measure the proton spectrum and the effective polar (vertical) source size of protons and X-rays. To this end, first, the particle beam is cut in half by the silicon knife edge positioned at L $= 45$–70 mm distance from the source. Knife edges were cut out from a 250 μm thick silicon wafer, with a length of 40–50 mm. They were positioned at a shallow angle (0.5–2°), in order to increase the effective path length (and thereby minimize transmission) for particle traces *through* the edge. After traversing the edge, the entrance slit (built of 40 mm thick steal plates) cuts a defined fan-beam out of the mixed particle beam. Right behind the entrance slit, the magnetic field bends charged particles away from the axis, depending on their charge-to-mass ratio and their energy. A layer of 1 mm thick CR39 in front of the IP detector stops ions heavier than protons in the relevant energy range, as well as potentially co-produced low-energy neutral particles, from the direct projection of the X-rays. The amount of magnetic deflection gives insight to the proton spectral distribution. Both in X-rays (direct projection) and protons (deflected signal),

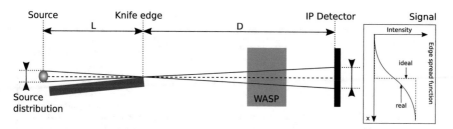

Fig. 6.2 Edge spread measurements. The top view schematic idea of the edge measurement and examples of the measurement outcome for an ideal (infinitely small point-like source) and for a realistic source with a finite distribution assuming geometrical optics

the edge-image is recorded (areas marked by dotted frames in Fig. 6.1) and encodes information on the effective source size in the direction perpendicular to the edge. The idea of this measurement and the projection geometry are sketched in more detail in Fig. 6.2. Measuring the edge-spread function (ESF) is a simple, inexpensive and widespread method to qualify the resolving power of optical systems [3]. We interpreted such measurement, similar to other measurements for X-ray [4] and neutron sources [5]. This setup involves no optical component. The measured ESF is thus a pure convolution of the source distribution and possible imperfections of the edge and the detector. When disregarding such imperfections, the "blur" of the recorded signal carries information on the source distribution (and effective source size) in the dimension perpendicular to the edge. Imperfections of edge and detector effectively make the measured source-size an upper limit, as they can add but not subtract from the true value.

The geometric magnification in this setup is $M = D/L$, where L is the source-edge distance and D is the edge-detector distance. In our experiments M took values of 17.5–27. Imaging plates were scanned at 25 μm nominal resolution with a calibrated General Electrics Typhoon FLA7000 scanner. The true resolving power is expected somewhat worse than that [6, 7]. We found realistic resolutions of 50–100 μm in edge-measurements with straight metallic edges (boundaries defined with better than 30 μm) mounted directly on top of the IP detector. This reduced real resolution stems from scattering of signal particles within the sensitive layer, and from optical scattering/blur in the optical readout procedure of the IP. The lower limit of reasonably observable source size given by the 27-fold magnification and the detector resolution is therefore in the 1.8–3.6 μm range. Alternatively, we replaced the edge-measurement at 83° with Aluminum filters of varying thickness ranging from 30 to 430 μm. These were mounted directly on top of the IP and allowed to infer an approximate X-ray spectrum according to [8], based on the X-ray attenuation.

At 103° in the horizontal plane, a similar combination of edge-image and WASP ($\mathcal{B} = 0.1$ T, $L_{\mathcal{B}} = 200$ mm, Drift $= 500$ mm) was employed to infer proton spectrum and azimuthal (horizontal) effective source size for X-rays and protons.

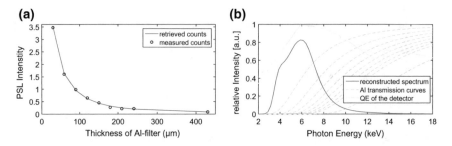

Fig. 6.3 X-ray spectrum. a IP signal in units of photostimulable luminescence (PSL), as measured behind different filter thicknesses and as retrieved by the algorithm. **b** IP response (light blue), Aluminum filter transmission (dashed) and retrieved X-ray spectrum (dark blue) for a single shot using a needle target. The reconstruction was done by Johannes Wenz, (group of Prof. Stefan Karsch, LMU)

6.1.1 X-Ray Spectrum

The X-ray spectrum was inferred from the filter stack measurement with the 83° diagnostics port. The recorded signal behind each filter is a convolution of IP response [9] and X-ray transmission through the filter [10]. Using methods detailed in [8] the X-ray spectrum shown in Fig. 6.3 was retrieved. The spectrum shows broadband yield up to the 10 keV level and is 'peaked' around 6 keV. We identify this with the Bremsstrahlung and recombination continuum content of typical spectra from laser-plasma interactions. The method can not resolve distinct peaks or discontinuities that may be produced in bound-bound or free-bound-emission, but can well serve to identify the approximate energy range of emitted photons and their usability for radiographic imaging in the observed energy range. The result is in general agreement with published results (e.g. [11, 12]). Low count numbers in low photon energies are due to constraints in the method; low energies will be heavily attenuated even in the thinnest aluminum filter and accordingly are set to zero in the retrieval.

A complementary approach was also used to estimate the spectrum from these filter measurements. The experimental data (Fig. 6.4a), with the known detector quantum efficiency and filter transmissions can be used to fit a spectral distribution of the form $dN/dE(E) = N_0 \cdot E^{-1} \cdot \exp(-E/k_B T_e)$, corresponding to Bremsstrahlung from a plasma [13] with electron temperature $k_B T_e$. Here, $E = h\nu$ is the photon energy and N_0 is a normalization factor. With a fit temperature of $k_B T_e = 2360 \pm 140$ eV (95% confidence level) and considering emission in the full solid angle, this spectrum contains an energy of 8.9 mJ, of which more than 3.8 mJ are emitted at energies higher than 2 keV. Reasonable photon numbers (e.g. at least 1 photon per pixel and keV) are observed at X-ray energies up to 10 keV.

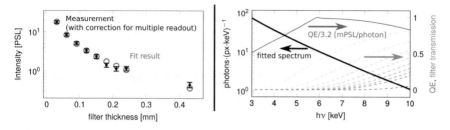

Fig. 6.4 Alternative retrieval of the X-ray spectrum. The fit shows good agreement with expectations from Bremsstrahlung

6.1.2 Proton Spectrum

A set of typical proton kinetic energy distributions recorded in a single laser-shot onto a needle target is displayed in Fig. 6.6. Alongside we show reference shots onto two types of foil targets, using 5 μm thick tungsten foil under $\alpha = 45°$ angle of incidence, and 190 nm thick Formvar (polymer) foil under $\alpha = 15°$ angle of incidence respectively. Angles are clarified in Fig. 6.5. The highest energy from the 190 nm thick Formvar [14] target, around 65 MeV, is recorded in laser propagation direction and comes close to currently achieved record energies (>85 MeV) at comparable laser-systems [15–17]. The needle target, similar to the thick foil target, reaches energies up to the 15–25 MeV level (depending on the observation angle). Both foil shots produce signal only in the diagnostic closest to the respective target normal. Contrasting this, the divergence of the beam generated in the needle shot is much wider due to the target geometry. Even more interestingly, spectral distributions observed towards both sidewards WASPs show distinctly peaked distributions around 12 MeV with FWHM of 2–3 MeV, and strongly resemble each other. Investigating more closely the proton acceleration from needle targets in Fig. 6.7 we observe, that experimental conditions including the laser pointing stability do strongly influence kinetic energy distributions, leading to shot to shot fluctuations in energy and particle number. The measurement of proton kinetic energy spectra in laser propagation direction via the TP is rather unspecific and possibly suffers from the high detection threshold in shots 30 and 33, due to the use of mm-thick CR39 filters on top of the IP. An important observations regarding both sidewards spectra must be highlighted: they resemble each other in single shots, showing comparable particle numbers in peaked spectra around 8–20 MeV with FWHM widths down to 2 MeV. The robustness suggests that ion-beam properties are indeed attributable to the microscopic nature of the target. It is known that a thin layer of protons on a metal substrate [18, 19] can lead to mono-energetic proton energy distributions. Heuristically, the protons accelerated from Tungsten needle targets originate from few-nm thick hydro-carbon contamination layers present on the target surface. They evince a massive positive charge buildup inside the original target volume, due to stripped electrons. From this charge buildup they are efficiently accelerated outwards by the Coulomb repulsion,

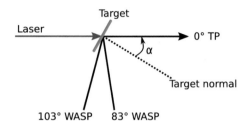

Fig. 6.5 Angle definitions used for foil targets

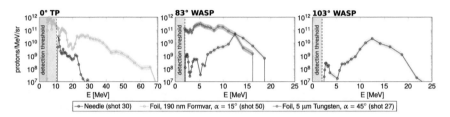

Fig. 6.6 Comparison of proton spectra for foil and needle targets. Single shot proton kinetic energy distributions measured by the respective diagnostics for a needle shot and two reference shots using foil targets. Faint color bands represent the numerical uncertainty of the detector evaluation and the separation of plot points along the energy axis indicates the spectrometer resolution

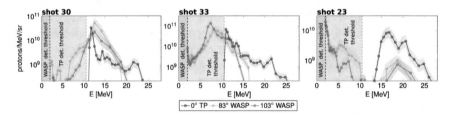

Fig. 6.7 Proton spectra for several shots at needles targets. Single shot proton kinetic energy distributions measured by the respective diagnostics for several needle shots. Faint color bands represent the numerical uncertainty of the detector evaluation and the separation of plot points along the energy axis indicates the spectrometer resolution

due to their high mobility. As they originate from an initially thin layer, and their acceleration is confined to small scales (e.g. they do not co-propagate with the laser pulse), the spatially limited extent translates to a limited spectral bandwidth [18, 19]. In addition, past work has observed (line-)focusing of the proton beam in laser-propagation direction by the self-induced magnetic fields in a wire target [20]. The spectral bandwidth comes to play a key role in imaging applications discussed later in this chapter; the heuristic hypothesis above concerning its origin is currently being tested in numerical studies, investigating the role of laser and target parameters in scalings of the proton peak energy and the spectral bandwidth, as well as the spatial distribution. This will hint future experimental directions.

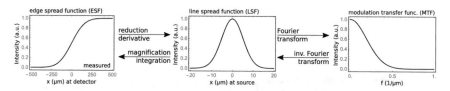

Fig. 6.8 ESF, LSF and MTF. The measured edge spread function (ESF) data, its relation to the 1D projection of the source intensity distribution (line spread function, LSF) and the modulation transfer function (MTF)

6.1.3 Effective Source Size

The measured ESF is related to the line-spread function (LSF) of the source via derivative and reduction (i.e. taking into account the magnification introduced by the projection geometry). The LSF is the 1D projection of the point spread function (2D intensity profile of the source). Its Fourier transform (the positive part thereof) is the modulation transfer function (MTF) often discussed to specify resolution, e.g. as the spatial frequency where the MTF yields an amplitude of 10%. These relations are displayed in Fig. 6.8.

For the presented evaluation we assumed a Gaussian LSF and fitted the corresponding distribution function (the error-function) to the recorded ESF data to retrieve the source-size. The combination of ESF measurements in polar/vertical and in azimuthal/horizontal direction yields a good idea of the underlying point spread function (PSF) of the source, and determines the achievable resolution in both directions, when the setup is used for imaging applications. Note that we refer to this measured source size as an 'effective' source size, which in principle could also be mimicked by a larger real source combined with a laminar flow, which could produce a similar signal [21].

Figure 6.9 shows an example of the recorded raw-data. Due to charged particles (e.g. electrons and faster protons) reaching the knife-edge before and while protons pass it, their traces are subject to energy dependent (read: time dependent) transverse deflections by induced electric fields. Such observation of electric fields by means of proton deflection is known as proton deflectometry mostly used to study actively (laser-)driven plasmas (e.g. [22–24]). However, the source remains confined for a given energy range (time). For this reason line-outs of only up to three lines (narrow energy/time band) are used to determine the proton source-size, contrasting the method for X-rays which uses the full slit-projection.

Exemplary X-ray measurements for a foil (5 μm thick Tungsten), and for both dimensions of the needle are displayed in Fig. 6.10. The foil shows a source size of 10.9 μm FWHM. The limited accuracy of the fit to the data in this case hints a more complex spatial distribution than a Gaussian, which may be expected from literature [25]. For either measured direction from needle targets, we observe smaller X-ray source sizes. On top of that, a difference arises between the dimension along the needle (polar) with 4.3 μm and the direction perpendicular to that (azimuthal) with

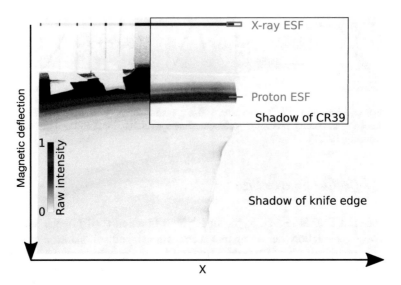

Fig. 6.9 ESF and spectrum raw data. A typical raw signal (recorded in shot 30 at $103°$) of the edge measurement, and indicators of the regions evaluated for the ESF. The complete IP is covered by 30 μm of Aluminum and partially covered by an additional 1 mm thick CR39 detector to clean proton signal from contaminants of heavier ions

2.4 μm FWHM source size, which is naturally expected from the target geometry. These small effective source sizes are within the lower limit of reasonably resolvable source-sizes of our method. I.e. true effective source-sizes could be even smaller.

In this case, the small source requires the correct treatment of the wave-nature of X-rays. In other words, the spatially coherent source requires the consideration of diffraction, when the X-rays are blocked in one half-plane by a knife-edge [4]. This diffraction from a knife-edge in the experimental configuration for a variation of source-size and including the spectral distribution (Fig. 6.3) can be calculated by Fresnel diffraction theory. For this purpose, the intensity on the screen position x is calculated as $I(x) = \mathcal{E}(x)\mathcal{E}^*(x)$. If the source is not point-like, the electric field is a convolution with the spatial source distribution $g(x)$, the spectral distribution $f(\lambda)$ and the spectral response function $R(\lambda)$ of the detector

$$\mathcal{E}(x) = \int_{x'} \int_\lambda g(x') f(\lambda) R(\lambda) \frac{1+i}{2} \left[\frac{1}{2} - \mathcal{C}(w(\lambda, x - x'D/L)) \right. \tag{6.1}$$
$$\left. -i \left(\frac{1}{2} - \mathcal{S}(w(\lambda, x - x'D/L)) \right) \right] dx' d\lambda,$$

with $\mathcal{S}$ and $\mathcal{C}$ representing the Fresnel integrals, and

$$w(\lambda, x) = x \frac{L}{L+D} \sqrt{\frac{2}{\lambda} \left(\frac{1}{L} + \frac{1}{D} \right)}. \tag{6.2}$$

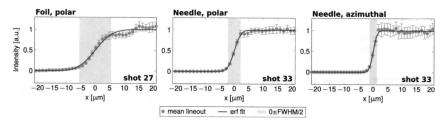

Fig. 6.10 Measured X-ray edge spread functions. Reference measurement for a foil target (in polar/vertical direction), and measurements for a needle target in both polar/vertical and azimuthal/horizontal directions. The source size is determined from a fit of the error-function to the data, i.e. assuming a Gaussian line spread function at the source. The FWHM source size retrieved from the fit is indicated by the green band

Such calculated intensity profiles are displayed in Fig. 6.11a. Here we used Gaussian intensity distributions of varying width, and the spectral distribution and spectral detector response information from Fig. 6.3. Three important insights are gained from there. First, the diffraction (e.g. the first maximum) occurs on a length scale close to 75 μm, which could be resolved by the detector. Second, the modulation amplitude for further maxima and minima decays very rapidly, which helps to increase the visibility of the first maximum, even in a detector with a limited resolution close to the width of the first diffraction maximum. Finally we observe that the diffraction is blurred, and loses visibility, already for a small finite source size. Data points of the azimuthal/horizontal measurement of shot 33 are shown alongside the calculations. The source-size of 2.4 μm FWHM retrieved earlier for this measurement, via fitting the error function, appears to be overestimated. In particular, the measured curve sits between calculated profiles for 1.15 and 2.35 μm FWHM source size. On the other hand, the absence of observable diffraction fringes in the data—even of the first and brightest one—provides a lower limit on the true source size of 1.15 μm. Indirectly, this measurement also confirms the spectral bandwidth of the source that was measured earlier (Fig. 6.3), since the diffraction in a narrower bandwidth source would be more pronounced in case of a small source size.

To generalize the diffraction-based discussion of the measured source-size, we note that a simple fit of the error-function to the calculated diffraction pattern tends to overestimate the source size, but quickly gains accuracy towards larger source size (Fig. 6.11b). This combined with the (missing) visibility of diffraction from measured signals, as a lower limit, provides reasonable results using the geometric optics approach (i.e. fit of the error function). Interestingly the small effective source size in X-rays enables phase-contrast imaging as demonstrated later in this chapter.

Similar measurements of the ESF for protons are presented in Fig. 6.12. The foil target shows a source-size of 8.4 μm FWHM, comparable to what was observed in X-rays. For needle targets the difference in both directions gets significantly more pronounced; for the polar/vertical dimension the source-size of 14.4 μm is even larger than that measured for a foil, while it is only 2.2 μm for the azimuthal/horizontal direction. Figure 6.13 wraps up these measurements, and demonstrates that described

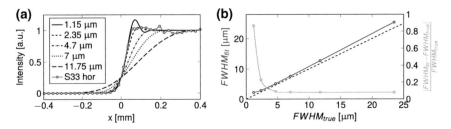

Fig. 6.11 Calculation of knife-edge diffraction. a Calculated intensity distributions in the detector plane for a variation of Gaussian source distributions (FWHM specified in the legend), shown together with measured data points for azimuthal direction in shot 33. **b** Comparison of FWHM source sizes used in the calculation and source-sizes retrieved from error-function fits to the resulting intensity distributions (left scale). The dashed line represents the (ideal) diagonal. The relative error is given on the right scale

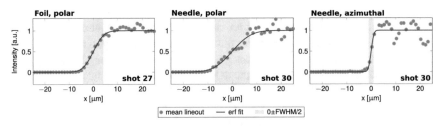

Fig. 6.12 Measured proton edge spread functions. Reference measurement for a foil target (in polar/vertical direction), and measurements for a needle target in both polar/vertical and azimuthal/horizontal directions. The source size is determined from a fit of the error-function to the data, i.e. assuming a Gaussian line spread function at the source. The FWHM source size retrieved from the fit is indicated by the green band

observations are consistent throughout the full dataset. It furthermore provides the error bars obtained from the 95% confidence level of the fit.

6.1.4 Discussion and Comparison to Alternative Sources

From the X-ray source distribution, the spectrum and the total signal recorded on the detector, we can derive some key-parameters of the source. In several shots, the X-ray image was saturated for the first IP scan. The corresponding PSL pixel count together with the IP response and our extracted spectrum yields an estimate of 2.8 photons/μm^2 at the detector position with the first retrieval method. The spectrum retrieved by the second method contains 0.8 photons/μm^2 above 1 keV. Assuming emission in the full solid angle and ps lifetime of the source [26], this amounts to a total energy in X-rays of 40 mJ and a power of 40 GW in a single shot (or 8.9 mJ and 8.9 GW using the alternative spectrum retrieval). We estimate the X-ray intensity at the needle-position to be around 10^{16} Wcm^{-2}, and with 100% bandwidth

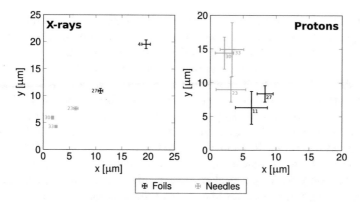

Fig. 6.13 Summary of source-size measurements. X-ray and proton source-size extracted from all measurements for the polar (y) and azimuthal (x) directions are displayed. For foils we measured only polar source distributions (y), due to the small beam divergence. Spherical symmetry was assumed as an estimate for the horizontal coordinate. In reality, the azimuthal source-size may be varied by additional geometric effects (angle of incidence and angle of projection as defined in Fig. 6.5) that do not occur in the polar source size. For needle-targets the proton source size was evaluated at the spectral peak. The proton source size for foils was evaluated close to the respective maximum energies. Small numbers close to the mark show the shot number. Error bars correspond to 95% confidence levels of the fit

the peak brilliance is about 10^{18} photons/mrad^2/mm^2/s/0.1%BW. These parameter values suffice to produce radiographic images of relatively large sized objects at a distance of 1 m from the source in a single laser shot.

The spectral and temporal characteristics of intense-laser-solid interaction driven X-ray sources have also been studied earlier (e.g. [11, 25, 27, 28]). Our results suggest, that towards our imaging-ports, X-rays are mostly emitted by laser-heated electrons falling back into the temporarily positively charged solid-density tungsten needle, acting as a micro-anode. Ultra-hot MeV-scale electrons that are generated by the laser ponderomotive force in such extreme intensities could provide larger conversion efficiency and photon energy, at cost of increased source size. Such electrons are however predominantly scattered and radiate along laser propagation direction (cf. Fig. 5.10). Meanwhile, the target-core consists of heavy ion species and stays relatively confined throughout the interaction; electrons falling back into that core lead to an intrinsically small source. Future investigations will look into these mechanisms of X-ray generation more closely, and evaluate possible scalings with laser parameters towards potential applications.

To set these values in perspective, we compare them to other common X-ray sources enabling inline phase contrast imaging. Modern free electron lasers like the linear coherent light source (LCLS) which deliver short (few fs), highly directed (mrad^2) beams of well defined photon energy, lead the field. A single LCLS pulse currently delivers energies in the mJ level at up to 12 keV photon energy. This produces 10^{33} photons/mrad^2/mm^2/s/0.1%BW peak brilliance (10^{21} to 10^{22} average at 120 Hz) mainly by virtue of a small bandwidth, small divergence and short pulse duration [29]. However, the source relies on a km scale particle accelerator. While the energy per shot in our setup does compete with an LCLS pulse, other parameters can naturally not hold up, due to the pulse duration, spatial spread, spectral bandwidth of the source and repetition rate. The accelerator-size demands more sensible comparisons of our X-ray source to smaller lab-scale sources which support in-line phase contrast imaging (PCI). Liquid metal jet micro X-ray tubes can provide high average power and small source-size (10 μm scale), but give no temporal resolution (e.g. [30, 31]). X-pinches produce small (μm-scale) high peak power (100 GW) sources that can be used for inline PCI with pulse duration of several tens of ps [32, 33]. The generation of sub-ps X-ray bursts from laser-solid interactions is established and has been used to study several physical systems mainly focused on temporal resolution rather than additional techniques like phase contrast imaging [27]. Lasers with mJ level pulse energy provide limited X-ray photon numbers and thus usually require multiple exposures to generate a reasonable image; an important caveat in case of sensitive samples that may be destroyed upon the first shot, or in case of instable dynamic systems that look different for each shot. Sources driven by short-pulse lasers in the 1–100 J range and solid-density targets often exhibit significant source size hindering single-shot PCI in a compact setup [11, 34, 35]. To minimize the spatial extent of the source, wires in the 10–20 μm range were used as targets, yielding source size of similar dimensions [28]. Such sources share the conversion efficiencies observed in our experiment in the 0.01–0.1% level, and their potential to be easily timed with respect to a high-power laser on a femtosecond level. Recently, efforts were made to exploit electrons accelerated in laser wakefield accelerators to produce synchrotron radiation with peak brilliances of 10^{22} photons/mrad^2/mm^2/s/0.1%BW [4, 36], again by virtue of small source size (μm^2), small beam divergence (mrad^2) and short pulse duration (fs). Concluding this discussion, solely based on X-rays, our source can be added to the range of capable laboratory-scale sources with sub-10-μm source size, albeit not unique in any of its properties.

What really sets this source apart from its competition is, that it produces energetic proton beams alongside the X-rays, which no other source including free electron lasers can currently provide. The laser-driven source from a foil target shows a typical decaying proton spectrum known from laser-driven sources for decades (cf. Ref. [37] and Fig. 6.6). From needles, we observe comparable spectra with lower maximum energies towards the laser propagation direction. Towards the sidewards ports we observe very atypical energy distributions for laser-driven accelerators. They can be referred to as quasi-monoenergetic in relation to laser-driven ion acceleration as they exhibit FWHM spectral bandwidths of only 2–3 MeV around peaks in the 8–20 MeV range ($\Delta E/E = 10$–37%), instead of the typical $\Delta E/E = 100\%$ for laser ion accelerators. Overall, this source shares the typical conversion efficiency from laser energy to accelerated protons in the percent range. The effective source size of our proton source, which amongst others determines the best achievable resolution in the desired radiographic image, shows an average asymmetry of 3 μm/13 μm, which mimics the elongated target geometry. Generally, it can in this respect very well compete with other laser-driven ion sources [21]. The visible asymmetry confirms the validity of the measurement method. Besides the source-size and the spectral distribution, an important figure for radiographic imaging is the particle flux at the positions of the sample and the detector; the peak in shot 30, considering its full width of 15 MeV, will disperse to a pulse duration of 8 ns at 0.5 m distance from the target. Due to the spatial divergence and the temporal dispersion, the flux depends heavily on the distance from the source. At 0.5 m distance, the spectral peak of shot 30 delivers about 8×10^6 protons/cm^2/ns (or 6.4×10^7 protons/cm^2/shot) towards the side ports.

The novel laser-driven proton source shall as well be compared to its alternatives in terms of key parameters. Conventional proton accelerators operating at comparable energies include the Tandem van-de-Graaf accelerator of the Maier-Leibnitz-Laboratory (MLL) in Garching, delivering protons up to 20 MeV in a narrow energy range ($\Delta E/E \approx 10^{-4}$) with proton flux of up to 10^9 protons/cm^2/ns delivered in ns bunch duration by means of special bunching techniques and within a focal area of $\sim 100 \times 100 \ \mu m^2$ [38] (10^5 protons/bunch, 16 μA). Particularly the bunch duration and the flux are outstanding features of this accelerator, right at the cutting edge of what is achievable with these machines. The MLL accelerator is about 40 m long and adds about the same length for several application beamlines, which clearly exceeds the footprint of the presented laser-driven setup.

The energy spread is much smaller in the conventional accelerator than in laser-driven accelerators, although the new needle-target has reduced the bandwidth of a laser driven source dramatically in comparison to a foil target. This existing but limited bandwidth comes to play a key role in the later radiographic imaging application. The particle flux of the MLL is far higher than that of our laser driven source measured at 0.5 m distance. This is at cost of a much reduced illuminated area, which requires stitching of areas for illumination of larger areas. The geometric dispersion of the laser-driven source intrinsically illuminates larger areas, and the beam can serve multiple purpose at once, e.g. assessing the spectrum while recording images of large objects simultaneously in a single laser shot. From a practical point of view, the use of CR39 detectors to record radiographic images demands detected particle flux below 10^8/cm^2 to avoid saturation[2] and thus benefits from the geometric dispersion; in imaging experiments we were able to adjust the CR39 detector position in the dispersing beam to approach the condition, while still coming close enough to the source with the X-ray detector (behind the CR39) to achieve significant signal to noise ratios for both.

Concluding the discussion, both the X-ray and the proton source alone are very capable to produce radiographic images in a single shot. Their combined use has never been demonstrated thus far. In the next section we will discuss the bi-modal radiographic imaging using such source.

---

[2]Considering that each proton leaves a track of at least 1 μm^2 diameter to be observable.

6.2   Simultaneous Bi-modal X-Ray and Proton Radiographic Imaging

Early efforts using laser driven sources for multi-modal emission and recording simultaneous X-ray and proton images, were limited to binary imaging of few-10-μm thin mesh objects [39, 40]. By using flat foil targets and comparably low power lasers by today's standards, the approach faced fundamental problems, which needed to be resolved for imaging thicker (more relevant) objects in more than a 'black-and-white' mode.

First, the proton kinetic energy used in [39, 40] was only up to 2–3 MeV, and distributed very broadly as typical in laser-plasma accelerators. This limited the thickness of objects (dynamic range), as well as the achievable contrast (if more than a binary mode/object would have been tried). Here, the proton energy has been increased beyond the 10 MeV level and features a broadly peaked spectrum.

Second, laser-plasma interactions with foil targets at increased laser power (such as used in this work to achieve higher kinetic proton energies), will lead to an increased source size. We confirmed this in reference measurements with foil targets and a PW laser, that are included in the source characterization above. In simple words, the area over which the laser can drive significant processes of emission is naturally larger when using a Petawatt instead of a Terawatt laser pulse. The proton beam can often maintain a small virtual source size due to the laminar flow of protons. However, the X-ray source is known to depend strongly on these effects. Here, this effect was demonstrated to be largely avoided by the use of micro-targets.

Last but not least, the higher particle energies are obtained by virtue of a higher laser peak-power. In order to still bring objects close enough to the source for imaging without object damage or image degradation (e.g., contribution or blurr by energetic electrons in the X-ray image), it is necessary to separate the desired particle beams for imaging (protons and X-rays) from:

- The transmitted laser pulse cone. Parts of the pulse will be transmitted or scattered in regimes that produce suitable ion beams, and then burn the object unless it is at a very significant distance.
- The very highly energetic electrons, which are mostly accelerated in the laser propagation cone (considering a thin foil or a needle), or along the target normal (considering a thick foil as used in [39, 40]). With PW lasers, those electrons can be very energetic (tens of MeV) and hard to magnetically deflect away from the X-ray beam over short distances. Hence, they would contribute to (blurr) the X-ray image.

Here, both the transmitted laser beam and the most energetic electrons are emitted primarily along the unaltered laser propagation direction cone. X-rays are expected to be emitted in the full solid angle and proton emission occurs around the target normal. This allows for a simple and effective geometrical separation of imaging beams (protons and X-rays) from laser and electron beams.

In the following, this section describes the use of both particle sources for radiographic imaging.

6.2.1   Some Ideas on Proton Radiography

Before jumping right into experimental approaches and results, a short introduction of the idea of quantitative proton radiography with a laser driven source will be given. The very basics of X-ray and proton interactions with matter are summarized in Appendix C.

1. Monoenergetic spectrum The monoenergetic ion beam with the localized energy deposition and correspondingly the localized ion-stopping position in matter (cf. Bragg peak, Fig. 1.3) represents the first and well studied scenario. Early trials using very monochromatic proton beams from conventional accelerators used exactly this property for ultra-high contrast proton-absorption-imaging of biological samples [41–43] by simply assessing the number/fraction of transmitted ions. The obvious disadvantage is, that contrast for a fixed proton energy is produced only for a small range of effective object thickness. Biological samples had to be immersed in solutions emulating the sample-stopping properties in order to produce reasonable images, i.e. to show contrast for density variations in the sample caused by diseases like tumors or strokes. Resulting images contained mostly binary information with the beam penetrating through—or not penetrating through—the sample in a certain lateral position.

  More recent approaches of proton imaging make use of conventionally accelerated higher energy protons, that easily penetrate through the sample and still carry significant residual energy. Intrinsically, such approach reduces dose on the sample (or patient). The energy is monitored before and behind the sample on a single ion basis. The simple two-dimensional particle-counting based transmission map of the sample from the early days is replaced by an energy-loss map. Consider an ion penetrating through a sample along a straight path in x-direction, with varying material properties along the path. It is useful to write the Bethe equation (C.10) as [44]

$$-\left\langle \frac{\mathrm{d}E}{\mathrm{d}x} \right\rangle = \varrho_x F(\iota_x, E_x) \simeq \varrho_x F_w(E_x). \tag{6.3}$$

Here, F contains the energy dependence and ϱ_x carries the local density information of the sample. The index x highlights quantities that depend on the locality (depth x) in the sample. In medical applications it is well established, that the mean ionization potential $\iota(x)$ is almost constant around 72–82 eV for human tissue and thus independent of x ($\iota_x \approx$ const.) [45]. Density and shell corrections are neglected. The energy dependence in F can then be expressed in terms of a known reference material, usually water (F_w), and variables in Eq. (6.3) can be separated and integrated [44]

$$T = -\int_{E_i}^{E_f} \frac{\mathrm{d}E}{F_w(E)} = \int_0^L \varrho_x \mathrm{d}x. \tag{6.4}$$

The value T can be identified with the water-equivalent thickness (WET) of the sample [46]. That is the thickness of a water sample that would cause the equivalent energy loss as the actual sample. It can be calculated from input and output energies via the middle part of the equation. The right hand side of the equation represents the Radon transform of the local relative stopping power ϱ_x and can be used in tomographic reconstruction when recorded from multiple projection angles [47]. The equation therefore represents the basis of modern proton CT. The valuable information here is the local relative stopping power ϱ_x, which can—once retrieved from a radiography or CT scan—directly be used in a treatment plan. Here we want to keep the particle counting approach and the 'instantaneous' image recording of all ions from earlier days while still retrieving information on the WET via the energy loss. To this end, we modify the input ion-beam spectrum.

2. Exponential spectrum (foil case) The second scenario is the typical laser-driven ion source. These exhibit very broadband spectral distributions with an approximately exponential shape (cf. Fig. 6.6a). The dynamic thickness range for which such spectrum can produce radiographic contrast is much larger than for the mono-energetic case. However, due to the slow variation in the spectrum, decent contrast in radiographic images is achieved only for significant changes in effective target thickness (Eq. (6.4)). Or in other words, small thickness variations will hardly vary the particle count in the detector. Due to the high particle count at low kinetic energies, a significant dose will be deposited inside the sample.

3. Limited bandwidth (needle case) The third scenario is the newly developed source, which features a limited spectral bandwidth. With such source, the dynamic range in terms of thickness is increased in comparison to the mono-energetic case but reduced compared to the foil scenario (exponential spectrum). Meanwhile, the contrast ratio of particle counts for variations in effective thickness is enhanced in comparison to the foil-case but reduced compared to the monoenergetic case. The dose deposited inside the sample is reduced as compared to the foil-case.

These three scenarios are wrapped up in examples displayed in Fig. 6.14. We assume initial kinetic energy distributions and evaluate their deformation by water samples of varying thickness, while keeping an eye on the detectable energy range. The first scenario (Fig. 6.14a) is a monoenergetic peak at 13.5 MeV. The visible finite width of the 'monoenergetic' beam reflects the finite resolution used for the calculation. The monoenergetic beam produces strong contrast for a very narrow dynamic range of effective sample thickness. In the second scenario (Fig. 6.14b) we approximate the typical laser-driven spectrum by an exponentially decaying spectrum $N(E) = N_0 \exp(-E/E_0)$ with characteristic energy E_0 of 10 MeV and a cut-off at 15 MeV. This scenario produces mediocre contrast for a broad dynamic range of effective sample thickness. The needle case (Fig. 6.14c) is modeled by a linearly increasing/decreasing spectrum with central energy of 10 MeV and full width of 10 MeV. This new scenario mixes both earlier cases, producing good contrast for a reasonable range of effective sample thickness. In the given example the contrast in terms of registered particle counts generated between water samples of thickness 0.7 and 0.9 mm is 0:1 for the monoenergetic beam, 1.4:1 for the foil case, and 1.9:1 for the needle case. This can benefit the imaging of biological and technical samples, where

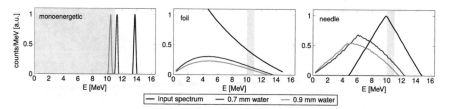

Fig. 6.14 Examples of proton radiography scenarios. Details of these scenarios are given in the text. The final spectrum behind samples of varying thickness was calculated via FLUKA [48, 49] simulations by Franz Englbrecht (at the chair of medical physics, Prof. Katia Parodi, LMU). Gray areas show the detected energy range in which particles are counted (corresponding to a broadband detector in the left, and thin CR39 detectors in the middle and right frame)

a limited dynamic thickness range (e.g. in a range of factor 2) is often acceptable as trade-off for enhanced image contrast. Considering the naturally existing detector noise and other uncertainties in 'particle counting', the contrast boost in comparison to the foil-scenario is equivalent to an increased effective thickness resolution, while keeping the dynamic thickness range reasonably large.

As long as the spectrum is known and monotone beyond the detected energy band, the counting based radiography principally allows quantitative retrieval of the effective thickness under the assumption of using a water sample (i.e. water equivalent thickness, WET), as a direct correlation of WET and particle count in the detector exists. This method is related to novel concepts of ion radiography using conventionally accelerated monoenergetic beams and stepwise scanning of the incident ion energy or use of a so called smeared out Bragg peak [50–52].

6.2.2 Setup

In order to record bi-modal radiographic images and simultaneously assess the input spectral information on the proton beam we use the setup depicted in Fig. 6.15a. The WASP at 103° is modified, increasing its slit-diameter to 2 cm and in this context referred to as X-ray cleaner (Appendix B.2.2). The proton image is recorded by the CR39 detector positioned directly before this X-ray cleaner at 0.49 m distance from the source. The front surface with a 30 μm aluminum filter is sensitive to 1.6–5 MeV protons. The backside is sensitive to 10.5–11.5 MeV protons, while heavier ions do not penetrate through the sample and the detector at present energy. The CR39 sensitive energy range was assessed by a witness CR39 detector that was used in the WASP and handled with the exact same procedures as the imaging CR39; it was additionally confirmed by SRIM calculations and is comparable to values from literature [53, 54]. Right after the recording of the proton image, the X-ray cleaner magnetically deflects any residual charged particles away from the direct projection of the widened slit. This projection is used to record the X-ray image with an IP at

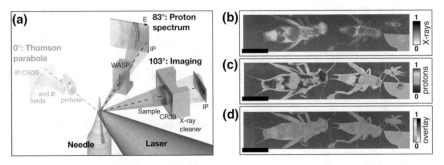

Fig. 6.15 Bimodal imaging of biological specimens. a Schematic of the setup for bi-modal imaging. Ions are indicated in green, X-rays are indicated in blue. The ion beam is monitored by the TP and 83° WASP. The sample is placed on a CR39 detector which registers the proton image. Behind the CR39 the X-ray cleaner deflects residual charged particles away from the direct projection and the IP records the X-ray image. **b** X-ray image of house crickets (acheta domestica, varying age/size) recorded in a single laser-shot on IP. **c** Proton image of crickets, recorded on CR39 in the same laser shot as b. The image was processed and recorded with a technique adapted from Refs. [55, 56] and records ion-impacts on the front- (1.6–5 MeV protons) and back (10.5–11.5 MeV protons) surfaces. **d** Overlay of both images, with the proton image scaled to 60% opacity. Scale bars in **b**, **c** and **d** correspond to 10 mm

0.75 m distance from the proton image. The magnifications in this setup are defined as

$$M = \frac{L + D}{L}. \tag{6.5}$$

Here L is the source-sample distance and D is the respective sample-detector distance, which differs for the proton and for the X-ray detector. For proton images presented here, $D = 0$ was used.

6.2.3 Results

A bi-modal radiograph of a series of biological samples (crickets) is depicted in Fig. 6.15. This demonstrates the basic capability to record relatively large objects in a single shot with both modalities. A bi-modal radiograph of a technological sample (smartphone camera adapter) is depicted in Fig. 6.16. Both X-rays and protons produce visible contrast over a variety of effective sample thickness. The discrepancy between both radiographs is likely caused by their different sensitivity to the atomic number in the material (Appendix C), and the presence of metals and polymers in the sample. After all, this is an important feature of the multi-modal imaging, as it shows that both images provide unique information. For a first quantitative analysis of the proton image, the spectrum measured at 83° is used to calculate the expected proton count in the detector for varying sample thickness under the assumption of imaging

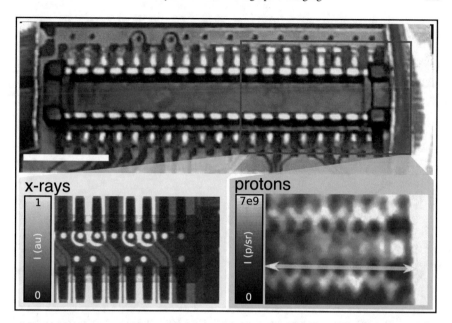

Fig. 6.16 Bimodal imaging of a technical sample. A simultaneous bi-modal proton and X-ray radiographic image of a technical sample (part of of a smartphone camera). In this case, the evaluation of the proton image was performed using a 20X dark-field microscope registering single proton impacts on the backside of the CR39 (10.5–11.5 MeV protons), allowing for quantitative analysis. The white scale-bar corresponds to 2 mm and is valid for all parts of the image

a water phantom[3] in Fig. 6.17a–b. It is then straightforward to find the thickness corresponding to the proton-count in a pixel, as shown in Fig. 6.17c–d.

At this point we want to discuss some possible measurement uncertainties of the method, and their impact on the result. The first form of uncertainty are absolute errors, which produce a general bias in the retrieved WET. There is no indication that this kind of error plays an important role in the current evaluation. The second kind of uncertainty are relative errors, which impact the calibration curve by a relative factor (i.e. the corresponding absolute error grows in relation to the particle count). Such uncertainty can be introduced by the calibration of the IP response to the proton dose, which is used to measure the spectrum. This relation is well known and linear in reality. Another error that mostly contributes in this form (assuming linear count/WET correlation) is the uncertainty in the determination of the detected energy range. Such errors lead to a variation of the slope of the calibration curve by a factor, and correspondingly a quenching or stretching of the entire curve in its height (in terms of WET). In the current version of this evaluation, we estimate these uncertainties to be of the order of 10% of the absolute values. The third uncertainty stems from statistical errors, mainly from the detector evaluation, which in the current

---

[3]By F. Englbrecht (at the chair of medical physics, Prof. Katia Parodi, LMU).

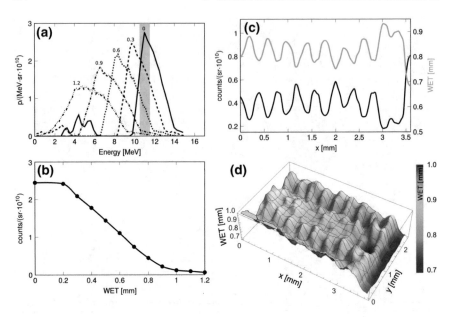

Fig. 6.17 Quantitative proton radiography. a A FLUKA [48, 49] simulation of the proton kinetic energy spectrum, modified by the passage of varying thickness of water. The thickness is specified in mm next to the spectra. The blue band shows the energy-band which is registered by the detector. The integration over this energy band leads to **b**. The calibration of detected proton count versus water equivalent thickness (WET). **c** Registered number of counts in the detector (black, left scale) in the line-out highlighted in Fig. 6.16 and retrieved WET of the imaged object (blue, right scale). **d** Full retrieved WET for the region in Fig. 6.16

version relies on particle counting. From the evaluation we estimate this uncertainty to be smaller than 10% of the absolute particle count. This corresponds to errors in WET of less than 0.05 mm and is consistent with the data, where we can observe variations of 0.1 mm WET. The last error stems from the spectrum measurement itself, or more precisely the limited correspondence between spectra measured at 83° and spectra used for imaging at the 103°. Different spectra lead to different calibration curves. This correspondence can amongst others be limited by the target surface roughness. In our experiments, the correspondence between spectra measured at both sideports (Fig. 6.7) stays mostly within the measurement accuracies of the spectrometers (typically factor 1.5 of the absolute signal in protons/MeV/sr). Future experimental work and numerical simulations will investigate this resemblance with complementary methods and higher resolution, allowing for its optimization. In the current work, this uncertainty is expected to contribute the largest part to overall uncertainties which may shift and/or quench the WET calibration by factors of 1.5.

Two more critical considerations concerning the physics need to be discussed for this evaluation technique: (a) how do the effects of lateral and longitudinal straggling influence the recorded image. And (b) how valid is the WET approach in an example, where metals are present in the sample (or what is its actual meaning in that case).

Question (a) shows as gradients in the proton image (and the retrieved WET) due to multiple Coulomb scattering, where the photograph and X-ray image show sharp edges. Using the estimate Eq. (C.18) for an incident pencil beam, a 1 mm thick object and and a 1 mm thick detector, predicts a Gaussian width of 70 μm in the image, which is about consistent with our measurements. This implies that gradients and particle counts in the image are correlated. In principle, this can be dealt with by iterative approaches. Such method would iteratively approach the true sample-WET distribution via forward-simulations, using the measured spectrum as input, the count-image as reconstruction goal, our first-order sample reconstruction as the starting point, and possibly the X-ray image as an additional input (e.g. to identify sharp edges in the object). A simple version of this idea was used in Ref. [21] for a different purpose. It would also intrinsically take into account the energy/range straggling. Note that energy and range straggling usually represent the fundamental limitation for the WET resolution (in methods detecting the residual energy of initially monochromatic ions behind and object via measurement by stopping them inside the detector and evaluating the Bragg peak). Hence, density/thickness resolution may benefit from the polychromatic approach (based on particle-counting in a single energy-bin) taken here, once that other measurement uncertainties are better controlled. In the current version the measurement uncertainty is dominated by other factors (see above).

Question (b) is difficult to answer in its generality. The water equivalence approach is at least established for most biological samples [46, 57].

Further information could be obtained about an object by detection of the image in multiple energy bands. First, that can be temporal resolution for imaging of dynamic systems, due to the run-time difference of different proton energies from the source to the sample [58]. Second, that can be further information about the final spectrum behind the sample, and therewith a more detailed idea of the sample composition. Finally, the consideration of multiple energy bands in combination with straggling processes may allow one to step away from the WET approximation and possibly retrieve even quasi-3D information from the sample in a single shot [59, 60].

Further examples of images obtained with the X-ray source at varying magnification (adjusted via the source-sample distance at constant source-detector distance) are given in Fig. 6.18 showing a cricket. Two important observations substantiate the source-size measurement presented earlier: first, the phase-contrast starts to show as edge-enhancements. Second, features corresponding to the source-size are resolved for sufficient magnification.

In many scenarios it is desirable to distinguish materials with comparable absorption coefficients, e.g. different kinds of soft tissue. Then, the radiographic contrast is often produced more efficiently using the wave-nature of X-rays. In phase-contrast imaging the radiographic contrast is produced by the phase shift of X-rays penetrating through the sample. Several schemes of PCI have been successfully tested, and many are nowadays used in research and early clinical trials [61–71]. For PCI in the point projection scheme as used here, the transverse coherence length of the source, geometrically defined as $d_c = \lambda L / \sigma$, should be equal or larger than the first Fresnel zone of radius $r_F = (\lambda D / M)^{0.5}$. Here λ denotes the X-ray wavelength, L denotes source-sample distance, σ denotes source-size and D denotes the sample-detector distance. In a point projection setup the magnification of the object on the detector is given by Eq. (6.5). The object-feature size a_M producing optimum phase-contrast can be found as $a_M = (2D\lambda / M)^{0.5}$, and thus detectors should feature sufficient resolution [72]

$$\Delta x \leq a_M M, \tag{6.6}$$

with Δx being the pixel size. The 2.56-fold magnification yields an absorption image because our detector cannot resolve the phase contrast according to the argument given in Eq. (6.6). Starting at 12.3-fold magnification the X-ray in-line phase-contrast becomes visible via edge-enhancements. With $\lambda = 0.2$ nm (corresponding to the 6 keV peak in the spectrum) we find for the required resolution $\Delta x < 74\,\mu$m, which is in accordance with the actual true detector resolution. Similarly, we find the condition $d_c \geq r_F$ verified ($d_c \approx 5\,\mu$m and $r_F \approx 5\,\mu$m).

Using even higher 21-fold magnification, features as small as 7.1 μm (peak to peak) and 3.5 μm (peak to valley) are observed in Fig. 6.18d–f in accordance with the measured source distribution measurement. The observed modulation depth is about 30% of the total signal. For a Gaussian LSF of the source, and for a object in the sample-plane of length scale 3.5 μm, such modulation depth corresponds to a maximum 12 μm (FWHM) source size. This can be found by comparison of the MTF for Gaussian distributions of varying widths, to the observed contrast in the measured image.

While the diffraction and the underlying geometry dictate the dominance of absorption or phase-contrast, the MTF ultimately limits the contrast attainable with either of them [72]. An additional 'damping' effect on the MTF, similar to the finite source size, is introduced with the finite spectral bandwidth of the source. Hence, the 12 μm effective source size represents the upper limit, which could possibly produce such contrast.

6.3 Conclusions

We have demonstrated a source that produces sufficient particle fluence at sufficient energies, both in X-rays and protons, to be usable as a single-shot backlighter for radiographic imaging. After characterizing these sources with respect to their spectral characteristics and spatial distributions, we recorded bi-modal images in a completely novel scheme. Outside of the laser-cone the sample can come close to the interaction without destruction by transmitted laser-light; this enables imaging and high magnification in a compact setup. By virtue of the unique spectral distribution observed in protons, we quantitatively retrieved the water-equivalent thickness of a sample. Despite possible uncertainties that arise, this is the first experiment and technique to our knowledge, showing quantitative acquisition of the two-dimensional

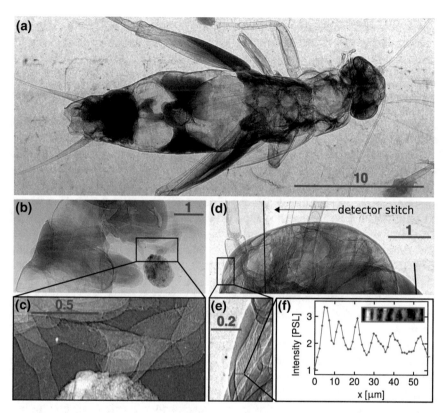

Fig. 6.18 X-ray images recorded at varying magnification. a Single shot 2.56-fold magnified absorption image of a cricket (acheta domesticus) that is 2.5 cm long in total. **b** Single shot 12.3-fold magnified image of cricket head/abdomen showing edge enhancement due to phase contrast. **c** Zoomed section from **b**. **d** Single shot 21-fold magnified image of a cricket head. **e** zoomed section of **d**. **f** Lineout (red) taken from **d**, inset showing the zoomed section. All scale bars are in units of mm

WET map of an object in a single laser-shot, i.e. within 10 ns. The presented example readily demonstrates the relative accuracy of the method. Future efforts using iterative approaches to reproduce the recorded image via simulations, and detection of the image in multiple energy bands, are likely to overcome current limitations.

The perfectly auto-registered X-ray image is recorded within the same laser-shot, i.e. only few ns before the proton image, from the very same original source. It could hence play an important role in the reconstruction procedure, providing the level of resolution and fine detail, which is missing from the proton image. The display of phase-contrast and demonstrated resolution (sub-5-μm) underline the potential.

Several possibilities exist to further increase the power of the presented methods. An increased energy of X-rays and protons would allow for imaging of thicker objects and seems within reach for next generation lasers, such as the ATLAS3000 currently being built at the Center for Advanced Laser Applications (CALA) in Garching.

Given the high repetition rate of such laser systems, the detection of bi-modal images will enable not only bi-modal radiography, but bi-modal tomography of biological samples. This could greatly benefit treatment planning in charged particle therapy of oncologic diseases, where as mentioned, the transfer from (typically) X-ray image to estimated proton stopping power is the largest error source [73, 74]. There are also possible variations of such setup, which could for instance magnetically focus protons to deliver therapeutic dose to a patient, while using the X-rays as an in-situ quality control.

It should be mentioned here, that not only the required laser-systems are on their way, but suitable targets and detectors are similarly under development. Besides trapped particles (cf. Chap. 5), suitable target concepts could be cryogenic jets [75], droplets [76], or even modified versions of target wheels [77] and tape-drive targets [78]. Promising detectors for such application need faster evaluation than those used here. A recent trend in our scientific community is the use of electronic or opto-electronic readout detectors [79]; in the imaging scheme we can use time-of-flight methods [80, 81], penetration-depth and detector-sensitivity methods as used here, or clever combinations [52, 82, 83] to retrieve energy distributions behind the object. The recording of initial energy distributions is a problem, which may be solved by the round target design and the resulting correlated spatio-spectral distributions of the source.

Apart from these 'real-world' applications, we foresee outstanding possibilities to diagnose ultrashort processes with the new bi-modal scheme. One obvious candidate are laser-plasmas, where diagnostics based on laser-driven X-rays [11, 84, 84] and protons [58, 85, 86] have a long standing history as a single species backlighter source. As protons react to density and to electric and magnetic fields present in a plasma, while X-rays react to density and material composition, their combination may give access to unambiguous information. Meanwhile, the energy distribution of protons may even provide some temporal resolution [58, 85, 86]. Of course, there are countless variations of such experiment, including (X-ray) pump (proton) probe experiments.

References

1. Bruker, www.bruker.com
2. Roth M et al (2002) Energetic ions generated by laser pulses: a detailed study on target properties. Phys Rev Spec Top Accel Beams 5(6):061301
3. Smith SW (1997) The scientist and engineer's guide to digital signal processing, 1st edn. California Technical Pub
4. Kneip S et al (2010) Bright spatially coherent synchrotron X-rays from a table-top source. Nat Phys 6(12):980983
5. Guler N et al (2016) Neutron imaging with the short-pulse laser driven neutron source at the Trident laser facility. J Appl Phys 120(15):154901
6. Izumi N et al (2006) Application of imaging plates to X-ray imaging and spectroscopy in laser plasma experiments (invited). Rev Sci Instrum 77(10):10E325

7. Boutoux G et al (2016) Validation of modelled imaging plates sensitivity to 1–100 keV X-rays and spatial resolution characterisation for diagnostics for the "PETawatt Aquitaine Laser". Rev Sci Instrum 87(4):043108

8. Sidky EY et al (2005) A robust method of X-ray source spectrum estimation from transmission measurements: demonstrated on computer simulated, scatter-free transmission data. J Appl Phys 97(12):124701

9. Meadowcroft AL, Bentley CD, Stott EN (2008) Evaluation of the sensitivity and fading characteristics of an image plate system for X-ray diagnostics. Rev Sci Instrum 79(11):113102

10. Henke BL, Gullikson EM, Davis JC (1993) X-ray interactions: photoabsorption, scattering, transmission, and reflection at E = 50–30000 eV, Z = 1–92. Atomic Data Nucl Data Tables 54(2):181–342

11. Park H-S et al (2008) High-resolution 17–75 keV backlighters for high energy density experiments. Phys Plasmas 15(7):072705

12. Yan J, Taewoo L, Rose-Petruck Christoph G (2003) Generation of ultrashort hard-X-ray pulses with tabletop laser systems at a 2-kHz repetition rate. J Opt Soc Am B 20(1):229–237

13. Huddlestone RH, Leonard SL (1965) Plasma diagnostic techniques. Academic Press

14. Formvar/Venylec. http://www.2spi.com/category/film-polyvinyl-vinylec/. Retrieved 29 Sept 2017

15. Wagner F et al (2016) Maximum proton energy above 85 MeV from the relativistic interaction of laser pulses with micrometer thick CH2 targets. Phys Rev Lett 116:205002

16. Jong Kim I et al (2016) Radiation pressure acceleration of protons to 93 MeV with circularly polarized petawatt laser pulses. Phys Plasmas 23(7):070701

17. Hegelich BM et al (Oct 2013) 160 MeV laser-accelerated protons from CH2 nanotargets for proton cancer therapy. ArXiv e-prints

18. Schwoerer H et al (2006) Laser-plasma acceleration of quasi-monoenergetic protons from microstructured targets. Nature 439(7075):445448

19. Robinson APL, Bell AR, Kingham RJ (2006) Effect of target composition on proton energy spectra in ultraintense laser-solid interactions. Phys Rev Lett 96:035005

20. Zulick C et al (2016) Target surface area effects on hot electron dynamics from high intensity laser-plasma interactions. New J Phys 18(6):063020

21. Borghesi M et al (2004) Multi-MeV proton source investigations in ultraintense laser-foil interactions. Phys Rev Lett 92(5):055003

22. Sokollik T et al (2009) Directional laser-driven ion acceleration from microspheres. Phys Rev Lett 103:135003

23. Borghesi M et al (2003) Measurement of highly transient electrical charging following high-intensity laser-solid interaction. Appl Phys Lett 82(10):1529–1531

24. Quinn K et al (2009) Laser-driven ultrafast field propagation on solid surfaces. Phys Rev Lett 102:194801

25. Faenov AY et al (2015) Nonlinear increase of X-ray intensities from thin foils irradiated with a 200 TW femtosecond laser. Sci Rep 5:13436

26. Murnane MM, Kapteyn HC, Falcone RW (1989) High-density plasmas produced by ultrafast laser pulses. Phys Rev Lett 62(2):155–158

27. Pfeifer T, Spielmann C, Gerber G (2006) Femtosecond X-ray science. Rep Prog Phys 69(2):443505

28. Brambrink E et al (2009) Direct density measurement of shock-compressed iron using hard X-rays generated by a short laser pulse. Phys Rev E 80(5):056407

29. Parameters of the linear coherent light source, Mar 2017

30. Larsson DH et al (2011) A 24 keV liquid-metal-jet X-ray source for biomedical applications. Rev Sci Instrum 82(12):123701

31. Otendal M et al (2008) A 9 keV electron-impact liquid-gallium-jet X-ray source. Rev Sci Instrum 79(1):016102

32. Shelkovenko T, Pikuz S, Hammer D (2015) X-pinches as broadband sources of X-rays for radiography. J Biomed Sci Eng 08(11):747–755

33. Zucchini F et al (2015) Characteristics of a molybdenum X-pinch X-ray source as a probe source for X-ray diffraction studies. Rev Sci Instrum 86(3):033507
34. Lancaster KL et al (2007) Measurements of energy transport patterns in solid density laser plasma interactions at intensities of 5×10^{20} W cm $^{-2}$. Phys Rev Lett 98(12):125002
35. Stephens RB et al (2004) K α fluorescence measurement of relativistic electron transport in the context of fast ignition. Phys Rev E 69(6):066414
36. Wenz J et al (2015) Quantitative X-ray phase-contrast microtomography from a compact laser-driven betatron source. Nat Commun 6:7568
37. Daido H, Nishiuchi M, Pirozhkov AS (2012) Review of laser-driven ion sources and their applications. Rep Prog Phys 75(5):056401
38. Dollinger G et al (2009) Nanosecond pulsed proton microbeam. Nucl Instrum Methods Phys Res Sect B Beam Interact Mater Atoms 267(12):2008–2012 Proceedings of the 11th International Conference on Nuclear Microprobe Technology and Applications and the 3rd International Workshop on Proton Beam Writing
39. Orimo S et al (2007) Simultaneous proton and X-ray imaging with femtosecond intense laser driven plasma source. Jpn J Appl Phys 46(9A):5853–5858
40. Nishiuchi M et al (2008) Laser-driven proton sources and their applications: femtosecond intense laser plasma driven simultaneous proton and X-ray imaging. In: Journal of physics: conference series, vol 112, p 042036
41. Steward VW, Koehler AM (1973) Proton radiographic detection of strokes. Nature 245(5419):38
42. Steward VW, Koehler AM (1973) Proton beam radiography in tumor detection. Science 179(4076):913–914
43. Koehler AM (1968) Proton radiography. Science 160(3825):303–304
44. Schulte R et al (2004) Conceptual design of a proton computed tomography system for applications in proton radiation therapy. IEEE Trans Nucl Sci 51(3):866–872
45. Besemer A, Paganetti H, Bednarz B (2013) The clinical impact of uncertainties in the mean excitation energy of human tissues during proton therapy. Phys Med Biol 58(4):887
46. Zhang R, Newhauser WD (2009) Calculation of water equivalent thickness of materials of arbitrary density, elemental composition and thickness in proton beam irradiation. Phys Med Biol 54(6):1383
47. Bushberg JT et al (2012) The essential physics of medical imaging, 3rd edn. Lippincott Williams and Wilkins, Philadelphia, PA
48. Böhlen TT et al (2014) The FLUKA code: developments and challenges for high energy and medical applications. Nucl Data Sheets 120:211–214
49. Ferrari A et al (2005) FLUKA: a multi particle transport code. CERN-2005-10, INFN/TC-05/11, SLAC-R-773
50. Telsemeyer J, Jäkel O, Martisikova M (2012) Quantitative carbon ion beam radiography and tomography with a at-panel detector. Phys Med Biol 57(23):7957
51. Ryu H et al (2008) Density and spatial resolutions of proton radiography using a range modulation technique. Phys Med Biol 53:5461–5468
52. Testa M et al (2013) Proton radiography and proton computed tomography based on time-resolved dose measurements. Phys Med Biol 58:8215–8233
53. Jeong TW et al (2017) CR-39 track detector for multi-MeV ion spectroscopy. Sci Rep 7(1):2152
54. Sinenian N et al (2011) The response of CR-39 nuclear track detector to 1–9 MeV protons. Rev Sci Instrum 82(10):103303
55. Gautier DC et al (2008) A simple apparatus for quick qualitative analysis of CR39 nuclear track detectors. Rev Sci Instrum 79(10):10E536
56. Paudel Y et al (2011) CR39 imaging technique for quick track analysis of particles generated in high-intensity laser target interactions. J Instrum 6(08):T08004–T08004
57. Newhauser WD, Zhang R (2015) The physics of proton therapy. Phys Med Biol 60(8):R155
58. Borghesi M et al (2001) Proton imaging: a diagnostic for inertial confinement fusion/fast ignitor studies. Plasma Phys Control Fusion 43(12A):A267–A276
59. Ruhl H Private communication

60. Raytchev M, Seco J (2013) Proton radiography in three dimensions: a proof of principle of a new technique. Med Phys 40(10). 101917, 101917-n/a
61. Davis TJ et al (1995) Phase-contrast imaging of weakly absorbing materials using hard X-rays. Nature 373(6515):595598
62. Wilkins SW et al (1996) Phase-contrast imaging using polychromatic hard X-rays. Nature 384(6607):335338
63. Franz P et al (2006) Phase retrieval and differential phase-contrast imaging with low-brilliance X-ray sources. Nat Phys 2(4):258–261
64. Cloetens P et al (1997) Observation of microstructure and damage in materials by phase sensitive radiography and tomography. J Appl Phys 81(9):5878
65. Diemoz PC et al (2013) X-ray phase-contrast imaging with nanoradian angular resolution. Phys Rev Lett 110:138105
66. Gureyev TE et al (2000) Quantitative methods in phase-contrast X-ray imaging. J Digit Imaging 13(1):121–126
67. Miao H et al (2015) Enhancing tabletop X-ray phase contrast imaging with nano-fabrication. Sci Rep 5:13581 EP
68. Houxun M et al (2016) A universal moire effect and application in X-ray phasecontrast imaging. Nat Phys 12(9):830–834
69. Langer M et al (2008) Quantitative comparison of direct phase retrieval algorithms in in-line phase tomography. Med Phys 35(10):4556–4566
70. David C et al (2002) Differential X-ray phase contrast imaging using a shearing interferometer. Appl Phys Lett 81(17):3287–3289
71. Munro PRT et al (2012) Phase and absorption retrieval using incoherent X-ray sources. Proc Natl Acad Sci 109(35):13922–13927
72. Mayo SC et al (2002) Quantitative X-ray projection microscopy: phase-contrast and multispectral imaging. J Microsc 207(Pt 2):79–96
73. Hanson KM et al (1982) Proton computed tomography of human specimens. Phys Med Biol 27(1):25
74. Hernáandez LM (2017) Low-dose ion-based transmission radiography and tomography for optimization of carbon ion-beam therapy. PhD thesis. Ludwig- Maximilians-Universität Müunchen
75. Göde S et al (2017) Relativistic electron streaming instabilities modulate proton beams accelerated in laser-plasma interactions. Phys Rev Lett 118:194801
76. Ter-Avetisyan S et al (2012) Generation of a quasi-monoergetic proton beam from laser-irradiated sub-micron droplets. Phys Plasmas 19(7):073112
77. Gao Y et al (2017) An automated, 0.5 Hz nano-foil target positioning system for intense laser plasma experiments. High Power Laser Sci Eng 5
78. Noaman-ul-Haq M et al (2017) Statistical analysis of laser driven protons using a high-repetition-rate tape drive target system. Phys Rev Accel Beams 20(4):041301
79. Reinhardt S (2012) Detection of laser-accelerated protons. PhD thesis. LMU
80. M Würl et al (2017) Experimental studies with two novel silicon detectors for the development of time-of-flight spectrometry of laser-accelerated proton beams. In: Journal of physics: conference series, vol 777, no 1, p 012018
81. Choi IW et al (2009) Absolute calibration of a time-of-flight spectrometer and imaging plate for the characterization of laser-accelerated protons. Meas Sci Technol 20(11):115112
82. Dromey B et al (2015) Picosecond metrology of laser-driven proton bursts. Nat Commun 7:10642
83. Dover NP et al (2017) Scintillator-based transverse proton beam profiler for laser-plasma ion sources. Rev Sci Instrum 88(7):073304
84. Brambrink E et al (2009) X-ray source studies for radiography of dense matter. Phys Plasmas 16(3):033101
85. Sokollik T et al (2008) Transient electric fields in laser plasmas observed by proton streak deflectometry. Appl Phys Lett 92(9):091503
86. Abicht F et al (2014) Tracing ultrafast dynamics of strong fields at plasma-vacuum interfaces with longitudinal proton probing. Appl Phys Lett 105(3):034101

Part IV
Summary and Perspectives

Chapter 7
Summary

In summary, we have studied the use of micro-plasmas as a source for ions and X-rays in a variety scenarios. Here, the major achievements of these efforts will be recapitulated and concluded in short terms.

In the first step, a novel Paul-trap-based target system was devised (Chap. 4), which allows reproducible trapping and target positioning with stable micron accuracy in a high-power laser focus in vacuum. This tool enabled us for the first time, to use truly isolated and well defined targets, spanning an unprecedented range in target size and mass (target diameters of few hundred nm to few ten μm). Thereby, the Paul trap closed the previously existing gap in target size, between other target technologies; those were either limited to the upper (e.g. droplet targets) or lower (e.g. gas cluster targets) limits of our target system. Correspondingly, a range of acceleration mechanisms (ambipolar, Coulomb explosion and directional) could now be accessed with a single target technology. A theoretical framework to estimate changes in the acceleration mechanism was developed (Sect. 5.1.3) and verified in simulations (Sect. 5.2) and experiments (Sects. 5.3 and 5.4). Now, the resultant ion-beams could be manipulated deterministically in terms of the spectral distribution by adjusting the target size to the laser pulse in order to vary the acceleration mechanism—short: (target)-size matters.

Importantly, a series of configurations was found that allowed to produce quasi-monoenergetic distributions with different degrees of directionality (and hence particle fluence). Figure 7.1 summarizes the results of these efforts and puts them in context with earlier experiments, highlighting the unprecedented kinetic energies and particle counts reached in *all* our experiments. This represents a major advance in the capability to generate narrow energy spread laser driven proton bunches. Already earlier in this work (Sect. 5.5) we have discussed some potential applications of such a source.

© Springer Nature Switzerland AG 2019
T. Ostermayr, *Relativistically Intense Laser–Microplasma Interactions*,
Springer Theses, https://doi.org/10.1007/978-3-030-22208-6_7

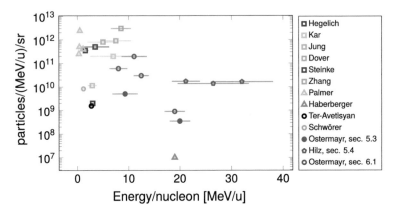

Fig. 7.1 Summary of experiments showing monenergetic ion spectrum. Our experiments in the context of earlier experimental efforts. Squares represent experiments with foil targets, triangles represent near-critical targets, circles represent mass-limited targets and pentagons represent near-critical mass-limited targets. Red markers represent work carried out in context with this thesis. Markers are positioned at the spectral peak, error bars represent the FWHM spectral bandwidth. To provide better readability, a selection of representative shots is shown. References: Hegelich [1], Kar [2], Jung [3], Dover [4], Steinke [5], Zhang [6], Palmer [7], Haberberger [8], Ter-Averisyan [9], Schwoerer [10]

The second pillar of this work concerns the development of one of these high-lighted applications. Using 'reduced dimension' tungsten micro-needles as targets, the strong X-ray emission (Sect. 6.1.1) and the quasi-monoenergetic ion energy distribution (Sect. 6.1.2) were measured. A small (few μm) effective source-size was found for both radiation modalities (Sect. 6.1.3). These spectral and spatial characteristics could be attributed to the reduced dimensions of the target, since tests with the same laser-pulse and bulk foil targets did—as expected—produce the typical ultra-broadband energy spectra and larger effective source-size at simultaneously reduced beam-divergence.

After this complete characterization, the unique beam qualities of the micro-needle target were implemented in the early development of a novel quantitative proton radiography based on the polychromatic proton source, in X-ray phase-contrast enhanced imaging showing few-μm resolution, and perhaps most importantly, the simultaneous radiographic imaging with both protons and X-rays from the very same source in a single laser-shot, i.e. within nanoseconds (Sect. 6.2).

Thus, microscopic targets have demonstrated tremendous potential for manipulation, optimization and application of the resulting particle and photon beams. Although suggested applications are in their infancy, using micro-targets has readily opened new perspectives for laser-driven sources.

References

1. Hegelich BM et al (2006) Laser acceleration of quasi-monoenergetic MeV ion beams. Nature 439:441
2. Kar S et al (2012) Ion acceleration in multispecies targets driven by intense laser radiation pressure. Phys Rev Lett 109(18)
3. Jung D et al (2011) Monoenergetic ion beam generation by driving ion solitary waves with circularly polarized laser light. Phys Rev Lett 107:115002
4. Dover NP et al (2016) Buffered high charge spectrally-peaked proton beams in the relativistic-transparency regime. New J Phys 18(1):013038
5. Steinke S et al (2013) Stable laser-ion acceleration in the light sail regime. Phys Rev Spec Top Accel Beams 16(1):011303
6. Zhang H et al (2017) Collisionless shock acceleration of high-flux quasimonoenergetic proton beams driven by circularly polarized laser pulses. Phys Rev Lett 119(16)
7. Palmer CAJ et al (2011) Monoenergetic proton beams accelerated by a radiation pressure driven shock. Phys Rev Lett 106(1):014801
8. Haberberger D et al (2012) Collisionless shocks in laser-produced plasma generate monoenergetic high-energy proton beams. Nat Phys 8:95–99
9. Ter-Avetisyan S et al (2012) Generation of a quasi-monoergetic proton beam from laser-irradiated sub-micron droplets. Phys Plasmas 19(7):073112
10. Schwoerer H et al (2006) Laser-plasma acceleration of quasi-monoenergetic protons from microstructured targets. Nature 439(7075):445–448

Chapter 8
Challenges and Perspectives

Several significant results were achieved within the framework of this thesis (previous section). Along the way, some challenges and opportunities were identified, which could benefit in one or another way from the presented developments. These perspectives will be discussed in this last chapter.

8.1 Laser Pointing Stability

A key problem throughout this thesis has been the limited success-rate of laser-shots, meaning the reduced probability of hitting a small target with a focused laser-pulse, due to laser pointing instabilities. In fact, laser pointing proves to be crucial, not only for experiments involving targets of reduced dimension. For instance, advanced laser wakefield accelerators require extremely stable pointing conditions when using capillary discharge waveguides, in order not to destroy them with a single shot.

Within this thesis, the damping of the harmonic motion of a levitating particle using opto-electronic feedback has been demonstrated. Assuming that most laser-pointing instabilities are caused by quasi-harmonic motions and vibrations inside the system with frequencies in the Hz (e.g. building vibrations) to kHz (e.g. vacuum pumps) range, we suggest to use a similar solution for pointing-damping of the actual high-power laser.

A possible implementation could use a continuous-wave guide-laser following the entire laser-chain up to the last turning mirror. The leakage of the guide-laser through that mirror will be used for a real-time far-field monitor using a PSD, tracking its position. In analogy to our Paul-trap damping, the PSD signal can be used to produce a damping signal. That can serve to drive fast servo-motors adjusting an early-stage turning mirror in the laser-chain, where the beam is still small, the required servo-travel is small and corresponding lever is large. This motorization will continuously correct for system-oscillations, such that full-system shots will in fact suffer

© Springer Nature Switzerland AG 2019
T. Ostermayr, *Relativistically Intense Laser–Microplasma Interactions*,
Springer Theses, https://doi.org/10.1007/978-3-030-22208-6_8

much less pointing inaccuracy. A comparable system has been operated at Lund University [1].

The reflection of the guide-laser from the leaky mirror will run further along the laser-beamline, including the focusing optics and the focus position. This provides further options to trigger the full-power shot exactly then, when the guide laser is *known* to be spot-on the target.

8.2 The Future of Trapped Targets

8.2.1 A Workhorse for Laser-Driven Accelerators

In the foreseeable future, the goal is to implement the Paul trap as target in the newly built 'Center for Advanced Laser Applications' (CALA) in Garching. This facility is explicitly aimed at the exploration of novel applications of laser-driven sources. The Paul trap [28] with the premier properties of ion bunches accelerated by its use could become an enabling technology to that end. In comparison to modern PW-class lasers, which usually cost more than 10^7 USD, the typical invest in target technology still lacks behind. In this broader context, the Paul trap is an approach to optimize the laser-target interaction as one of just two key-ingredients that can be tweaked.

Besides the ion-bunch properties, this target brings several practical advantages over other—and especially over bulk foil—targets. Targets are commercially available in large quantities, and come in monodisperse, chemically grown and well characterized samples. This reduces both cost and effort for on-site target-characterization (usually required for modern foil-targets [2]) to essentially zero. A single target can be as cheap as 10^{-10} USD. Another critical issue is the debris usually generated in the interaction. Much of this debris stems from the target-holder and outer regions of the target, where a laser-beam with peak intensity beyond 10^{18} Wcm^{-2} sill carries sufficient energy to ablate material. Debris degrades nearby optics and severely limits their lifetime (cf. Fig. 8.1). This is critical for real-world applications driven by laser-plasma accelerators, since especially the final focusing optic is unavoidably located close to the target, and accounts for a very significant part of costs of the complete target-area setup. Reducing the target to its bare minimum, and avoiding any excess mass, our target represents the natural optimization to reduce such problem.

Similarly, the reduced target-mass may significantly reduce the electro-magnetic pulse (EMP) generated during the interaction, due to electrons being accelerated and due to consequent neutralizing currents into the target. The reduction of EMP eases the operation of electronic online detectors, which often face difficulties in such environment. The use of electronic detectors (including charge-coupled devices for that matter) seems unavoidable for operation at relevant repetition rates (e.g. 1 Hz).

Fig. 8.1 Off-axis parabolic mirror degraded from laser and debris. The photograph shows a f/2.5 off-axis parabolic mirror (152 mm diameter) used at the Laboratory for Extreme Photonics (LEX) to focus the 2 J, 30 fs ATLAS laser onto solid density targets in a one-year long experimental campaign (few thousand shots)

Repetition Rate

In order to make full use of lasers with repetition rates around 1 Hz, an improvement of the current maximum repetition rate of 1 trapping process (and laser-shot) per 15 min is required.[1] This is pursued via automation of the complete initial trapping sequence. In the present state of the system, there are three manual steps required for the initial trapping of a particle, slowing down the overall performance. These include the power on/off of the ion gun, together with the gas-flow adjustment through it, the actuation of the particle reservoir mechanism, and the adjustment of end-cap voltages after the initial trapping process. As the experiment areas are usually controlled for radiation safety, any such manual interaction directly at the system costs time to re-establish the safety protocol. All these actions can be automated similar to the other parts of the system. Then, a single push-button will remotely trigger the full trapping sequence.

Once this development has been completed, the system will support much larger repetition rates than the current 1 shot per 15 min. The first cornerstone for this is the imaging system, which practically requires the adjustment of laser-target overlap

---

[1]This section is reproduced with small variations, and with permission, from the original peer-reviewed article: T.M. Ostermayr et al., Review of Scientific Instruments, 89:013302, (2018). The article is published by the American Institute of Physics and licensed under a Creative Commons Attribution 4.0 International License (https://creativecommons.org/licenses/by/4.0/).

only once by use of the additional focus-diagnostics. Once this overlap is established, the reproducible positioning of targets in the focus can be achieved equally well by means of the Paul trap optical system alone. But unlike the focus-diagnostics, this setup does not require physical motion of any optical element before and after the laser engagement, in order to bring it to a position save from the laser. This motion of an optical element over significant distances (several cm) at high speed and at high precision (μm) is difficult (if not impossible) to achieve at 1 Hz rate. The second cornerstone is the demonstrated reproducibility of charge-to-mass ratio for consecutively trapped particles. This allows to operate the system at constant parameters for a given choice of particle-sample, thereby reducing time and effort for consecutive trapping processes. The third and last cornerstone is the implementation of a hybrid damping-scheme combining buffer-gas damping for the early stage (shortly after trapping), and active damping for the final damping. This can reduce overall damping time considerably, since the currently deployed active damping is rather ineffective in the early stage, i.e. with initial particle trajectories exceeding the field of view of the optical system.

With the current modus operandi (trapping one particle at a time and staying in the weak damping approximation) being optimized for rep-rate, we anticipate an upper limit of repetition rate around 1 Hz. This is because we typically require hundreds of oscillation periods to damp the particle trajectory, and secular frequencies are typically in the few-hundred Hz range. Similarly, charging times will realistically stay in the second range. This would suffice to make full use the 1 Hz, 3 PW ATLAS3000 laser at CALA. Further increased rep-rates might be achievable by implementation of additional techniques, e.g. a segmented trap allowing to trap, charge and pre-damp particles in a preparatory trap segment, before injecting and shooting them in the target-trap segment. Another way could be an increased damping amplification to approach critical damping, which is however non-trivial in practice, e.g. due to the coupled measurement and damping of coordinates.

8.2.2 Novel Plasma Parameters

In addition to its use as a daily driver in laser-plasma accelerators, we consider the Paul trap as a unique tool to study and identify ideal target parameters at a given laser system and for a given purpose from the very broad supported target-material and size-range.[2] This allows one to invest reasonably into a more narrow-range solution developed specifically aiming at very high repetition rates ($\gg$1 Hz, e.g. droplet targets or cluster targets), or to develop new solutions, if the desired parameter-set is unavailable by other high-rep means. The necessity and requirements of such

---

[2]This section is reproduced with small variations, and with permission, from the original peer-reviewed article: T.M. Ostermayr et al., Review of Scientific Instruments, 89:013302, (2018). The article is published by the American Institute of Physics and licensed under a Creative Commons Attribution 4.0 International License (https://creativecommons.org/licenses/by/4.0/).

investment (or development) into narrow parameter-range targets will ultimately be driven and dictated by the aimed-at real-world application.

Besides laser-plasma accelerators, the related research of warm dense matter (WDM) and the isochoric heating of material can greatly benefit from isolated targets. Here, so-called reduced mass targets (RMT), held by tiny supporting structures, are often used for better comparability of experiments with simulations and analytical models [3–6]. In this field, similar to the laser-driven acceleration of particles, material close to the target, including the target-mount, neighboring targets and background gas, contribute to the laser-target interaction, e.g. via energy dissipation [7–9]. Target mounts, albeit tiny in scale and mass, can have profound implications even for the mm-scale plasmas, such as used in inertial confinement fusion [10].

Isolated Nanofoil-Targets

One particularly interesting scenario that has not been realized yet by any other method, are isolated foil-targets with diameters in the range of the laser focus size (or smaller), and thickness in the nm range. Simulations demonstrate that these targets hold great potential for for x-ray and ion beam generation [11–13]. The trap in principle allows to deliver such targets [14]. In fact, we have trapped carbon platelets (diameter 10 μm, thickness 300 nm) and aligned them with respect to a reference, by spinning them up with a small (50 mW, 660 nm) circularly polarized continuous wave laser. The transfer of angular momentum from circularly polarized light to absorbing objects is a well known effect [15–17]. Figure 8.2a–d illustrates the trapped carbon platelet recorded by the emulated focus diagnostics microscope, and backlighted by a 532 nm, 5 mW laser, without the circularly polarized laser engaged. Both the particle rotation and motion in the trap are observed. Figure 8.2e–h shows the same particle just after turning on the circularly polarized laser, still flipping and moving in the trap. Figure 8.2i–l shows the same particle after 1 min being illuminated by the circularly polarized laser. Both the positioning and the particle's orientation are stabilized, even without application of the active damping mechanism, in analogy to a spin-top. The size-variance in available target-samples has thus far limited the trap repetition rate to rather unpractical 1 per hour and is currently being investigated, together with details of the spin-up and spin-down behavior and the alignment, which itself may be of relevance for a variety of reasons [e.g. Refs. 14, 18].

Circular Laser Polarization

The circular laser polarization is a concept that could also proof interesting for the driving laser pulse in the laser-plasma interaction. Circular polarization of the originally linearly polarized laser pulse can be achieved by insertion of a $\lambda/4$-plate. In our past experiments using micro-targets, the circular polarization was unavailable. This may change in the near future, and is expected to further facilitate directional acceleration at simultaneously suppressed electron heating, increasing the overall efficiency (and ion energy). This is due to the constant push exerted by the ponderomotive force of a such polarized laser, which is in contrast to the 2ω shaking in the linearly polarized laser field. The relevance of this concept for bulk foil targets has been demonstrated by ours and other groups [19–21].

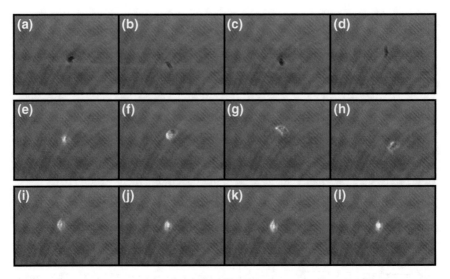

Fig. 8.2 Trapped and aligned carbon platelet. **a–d** Without circular laser illumination, recorded at temporal steps of 60 ms—not aligned. **e–h** Right after turning on the circ. pol. laser, still with random orientation. **i–l** After one minute of illumination with circ. pol. laser, aligned and stabilized in position. For reference, each image frame is 80 μm in height

8.3 Laser-Driven Micro-sources for Imaging

Of course, one important motivation for our work towards novel targets in laser-driven accelerators are the potential applications. Collaborations with the chair of medical physics at LMU are starting to explore, how to best use and improve the combined single-shot multimode and polychromatic image from a single source in medical physics. Meanwhile, ideas are also pursued to investigate fast processes by means of the multi-mode source, including warm dense matter and laboratory astrophysics. Novel particle-in-cell simulations investigate the relevant parameters that need to be optimized, in order to further increase the proton and X-ray energy for applications. The issue of fast and robust detection techniques is crucial not only to imaging schemes, but to laser-plasma accelerators in general, which is reflected in several new developments [22–27].

Last but not least, the acceleration of dense sub-fs relativistic electron bunches (cf. Fig. 5.10 in Sect. 5.2.1 and Refs. [29, 30]) may facilitate the revival of the relativistic mirror [31] and other concepts [32], to produce bright and short photon bursts. Both the electrons and potential tertiary sources promise imaging applications at unprecedentedly high spatial and temporal resolutions (nm, sub-fs) in a lab-scale setup.

References

1. Genoud G, et al (2011) Active control of the pointing of a multi-terawatt laser. Rev Sci Instrum 82(3):033102
2. Gao Y, et al (2017) An automated, 0.5 Hz nano-foil target positioning system for intense laser plasma experiments. High Power Laser Sci Eng 5
3. Myatt J et al (2007) High-intensity laser interactions with mass-limited solid targets and implications for fast-ignition experiments on OMEGA EP. Phys Plasmas 14(5):056301
4. Nishimura H, et al (2011) Energy transport and isochoric heating of a low-Z, reduced-mass target irradiated with a high intensity laser pulse. Phys Plasmas 18(2):022702
5. Neumayer P et al (2009) Isochoric heating of reduced mass targets by ultra-intense laser produced relativistic electrons. High Energy Density Phys 5(4):244–244
6. Theobald W, et al (2006) Hot surface ionic line emission and cold K-inner shell emission from petawatt-laser-irradiated Cu foil targets. Phys Plasmas 13(4):043102
7. Ter-Avetisyan S et al (2012) Generation of a quasi-monoergetic proton beam from laser-irradiated sub-micron droplets. English. Phys Plasmas 19(7):073112
8. Zeil K, et al (2014) Robust energy enhancement of ultrashort pulse laser accelerated protons from reduced mass targets. Plasma Phys Control Fusion 56:084004
9. Morace A, et al (2013) Improved laser-to-proton conversion efficiency in isolated reduced mass targets. Appl Phys Lett 103(5):054102
10. Nagel SR, et al (2015) Effect of the mounting membrane on shape in inertial cofinement fusion implosions. Phys Plasmas 22(2):022704
11. Bulanov SS, et al (200) Accelerating monoenergetic protons from ultrathin foils by at-top laser pulses in the directed-Coulomb-explosion regime. Phys Rev E 78:026412
12. Yu TP, et al (2014) Dynamics of laser mass-limited foil interaction at ultra-high laser intensities. Phys Plasmas 21(5):053105
13. Yu T-P, et al (2013) Bright betatronlike x rays from radiation pressure acceleration of a mass-limited foil target. Phys Rev Lett 110:045001
14. Kane BE (2010) Levitated spinning graphene flakes in an electric quadrupole ion trap. English. Phys Rev B 82(11):115441
15. Friese MEJ et al (1996) Optical angular-momentum transfer to trapped absorbing particles. Phys Rev A 54(2):1593–1593
16. Friese MEJ et al (1998) Optical alignment and spinning of laser trapped microscopic particles. Nature 394:348–350
17. Beth RA (1936) Mechanical detection and measurement of the angular momentum of light. Phys Rev 50 :115–125
18. Abbas MM et al (2004) Laboratory experiments on rotation and alignment of the analogs of interstellar dust grains by radiation. Astrophys J 614:781–795
19. Kar S, et al (2012) Ion acceleration in multispecies targets driven by intense laser radiation pressure. Phys Rev Lett 109(18)
20. Jung D, et al (2011) Monoenergetic ion beam generation by driving ion solitary waves with circularly polarized laser light. Phys Rev Lett 107:115002
21. Steinke S, et al (2013) Stable laser-ion acceleration in the light sail regime. English. Phys Rev Spec Top Accel Beams 16(1):011303
22. Reinhardt S (2012) Detection of laser-accelerated protons. PhD thesis, LMU
23. Würl M, et al (2017) Experimental studies with two novel silicon detectors for the development of time-of-ight spectrometry of laser-accelerated proton beams. J Phys: Conf Ser 777(1):012018
24. Choi IW, et al (2009) Absolute calibration of a time-of-flight spectrometer and imaging plate for the characterization of laser-accelerated protons. Meas Sci Technol 20(11):115112
25. Dromey B, et al (2015) Picosecond metrology of laser-driven proton bursts. Nat Commun 7:10642. Challenges Perspect 160(8)
26. Dover NP, et al (2017) Scintillator-based transverse proton beam profiler for laser-plasma ion sources. Rev Sci Instrum 88(7):073304

27. Haffa D, et al (2017) Ion bunch energy acoustic tracing (I-BEAT). Submitted manuscript
28. Ostermayr TM, et al (2018) A transportable Paul-trap for levitation and accurate positioning of micron-scale particles in vacuum for laser-plasma experiments. Rev Sci Instrum 89:013302
29. Cardenas DE (2017) PhD thesis, Ludwig-Maximilians-Universität München
30. Cardenas DE, et al (2017) Relativistic nanophotonics in the sub-cycle regime. Submitted manuscript
31. Kiefer D, et al (2013) Relativistic electron mirrors from nanoscale foils for coherent frequency upshift to the extreme ultraviolet. Nat Commun 4:1763
32. Ma WJ, et al (2014) Bright subcycle extreme ultraviolet bursts from a single dense relativistic electron sheet. Phys Rev Lett 113(23)

Appendix A
Texas Petawatt Laser

The following schematics show the stretcher, the amplifiers and the compressor designs of the Texas Petawatt laser. They are reproduced from the official Texas Petawatt website http://texaspetawatt.ph.utexas.edu/overview.php, retrieved 07. Nov. 2017, where larger scale and up-to-date versions are provided.

© Springer Nature Switzerland AG 2019   147
T. Ostermayr, *Relativistically Intense Laser–Microplasma Interactions*,
Springer Theses, https://doi.org/10.1007/978-3-030-22208-6

Stretcher

Path Through Stretcher

1. 1" Mirror (1)
2. Grating (below striped mirror)
3. Curved Mirror ⎤
4. Striped Mirror ⎬ Telescope
5. Flat Mirror ⎦
6. Striped Mirror ⎤
7. Curved Mirror ⎦
8. Grating (above striped mirror)
9. Roof Top Mirror
10. Grating (above striped mirror)
11. Curved Mirror ⎤
12. Striped Mirror ⎬ Telescope
13. Flat Mirror ⎦
14. Striped Mirror ⎤
15. Curved Mirror ⎦
16. Grating (below striped mirror)
17. Lens ⎤
18. 1" Mirror (2) ⎬ Telescope
19. 1" Mirror (3) ⎦

Beam then retraces steps back through
stretcher, through the Faraday Rotator
and is eventually ejected off a polarizer.

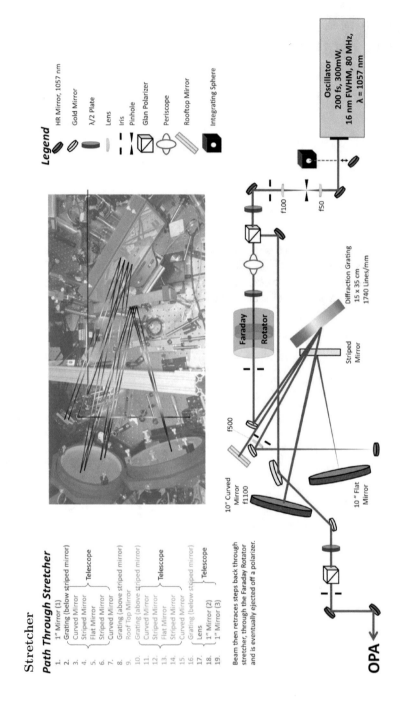

Legend

- HR Mirror, 1057 nm
- Gold Mirror
- λ/2 Plate
- Lens
- Iris
- Pinhole
- Glan Polarizer
- Periscope
- Rooftop Mirror
- Integrating Sphere

Oscillator
200 fs, 300mW,
16 nm FWHM, 80 MHz,
λ = 1057 nm

f100

f50

Faraday Rotator

Diffraction Grating
15 x 35 cm
1740 Lines/mm

Striped Mirror

f500

10" Curved Mirror
f1100

10" Flat Mirror

OPA

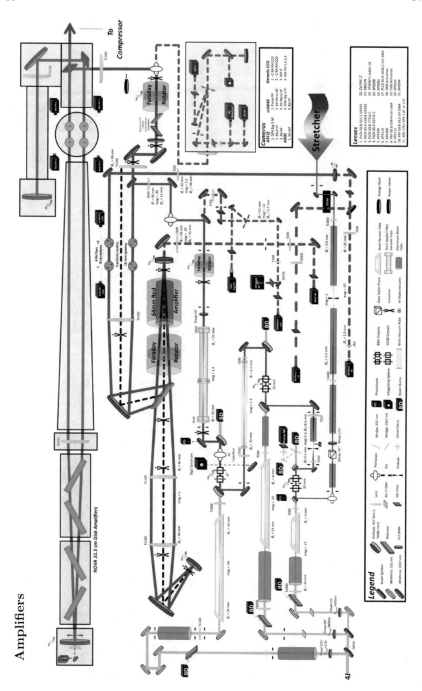

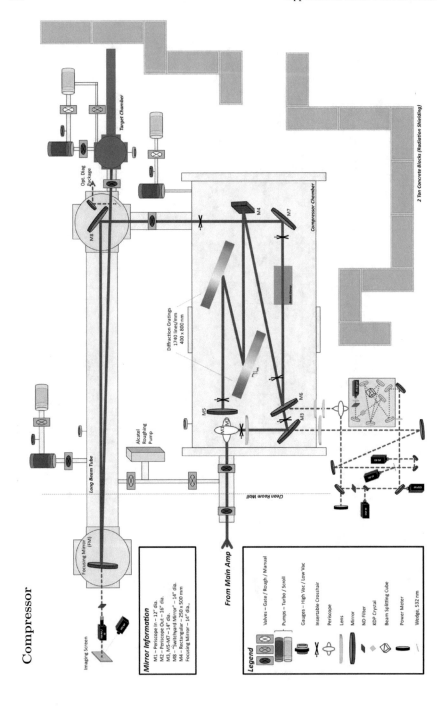

Appendix B
Diagnostics

B.1 Detectors

B.1.1 CR39

The CR39 detector (Coulumbia resin #39) is a polymer with the chemical composition $C_{12}H_{18}O_7$ and a mass-density of 1.3 g cm^{-3}. In our experiments we use the commercial type of CR39 (TASTARK, Track Analysis Systems Ltd.) as an area solid state nuclear track detector which comes as a optically transparent plastic plate (the ones used here had dimensions of ~50 × 50 × 1 mm). An ion penetrating into such a plate deposits energy and ionizes material along its path. Chemical bonds can be broken if the energy deposited exceeds the material-specific threshold. This leaves a track of broken bonds along the particle-trajectory through the material. Due to the material-threshold for ionization, the CR39 detector is sensitive exclusively to ions.

In order to visualize the track created by an ion, the detector plate is etched in caustic alkaline solution (NaOH, 6 mol/L, 80 °C). The etching time for data presented here was varied between 30–50 min. The difference in the etching rates between damaged and undamaged regions of CR39 results in the formation of conical holes in the surface of CR39 at positions where ions were impacting. The exact size and shape of these 'pits' can further serve to distinguish protons from heavier ion species, and potentially to recover spectroscopic information. In order to quantify and visualize the information stored on a detector, its surface is recorded using an automated dark-field microscope. From these images, the position, size and shape of the impact can be extracted and used for further analysis. Of course, such a analysis is time consuming but allows the quantitative treatment of the data (e.g. particle counting).

Alternatively, in order to more rapidly record large-scale images after the etching process, a laser was coupled into the transparent detector plate through its edge. A certain portion of the light reflects internally and stays trapped within the detector. Similar to the dark-field microscopy method, the small holes produced by the ions will stray the light, which can be recorded with a suitable camera setup [3, 4]. This approach has not been engineered towards quantitative analysis by anyone thus far,

© Springer Nature Switzerland AG 2019
T. Ostermayr, *Relativistically Intense Laser–Microplasma Interactions*,
Springer Theses, https://doi.org/10.1007/978-3-030-22208-6

which however seems feasible. It does however give signal for both the front- and backside of the detector.

Information on the particle energy can be gained by means of the penetration depths of respective proton-energies in combination with SRIM simulations and/or measurements regarding the visible energy range on the detector when placing the detector in a magnetic spectrometer. For protons that is $\sim 0.1 - 5$ MeV on the front surface and 10.5–11.5 MeV on the rear surface for data presented herein.

In the scope of this thesis, CR39 detectors have also been used as a cross-calibration for data recorded by other means (imaging plates), as well as a stack-detector for measurements of proton energies. Its exclusive sensitivity to ions was exploited to record bi-modal ion- and X-ray-images in a single-shot setup.

B.1.2 Imaging Plate

The imaging plate (IP) is an area detector which uses the photo stimulated luminescence of phosphor (PSL) to record, store and analyze the signal. The BAS-TR type imaging plate by Fuji film, as used in this work, consists of three layers: a (protection-less) 50 μm thick phosphorus layer ($BaFBr_{0.85}I_{0.15}$), a 250 μm thick polymeric support layer and a 160 μm thick magnetic layer.

When ionizing radiation interacts with the sensitive phosphorus layer, charge carriers can be trapped in a meta-stable state of the phosphorus and thereby store a latent image of the incident radiation in the IP. A small amount of trapped charge escapes with time, and thus the signal fades characteristically. In this work, we evaluated all IP at 15–20 min after the recording of data. The signal analysis takes place in the imaging plate scanner. For this work we used scanners by Fuji Film and General Electrics, types FLA5100 and FLA7000. In the scanner, energy is transferred to the trapped charges via photons (red laser, 633 nm, HeNe), to elevate them to the conduction band. Upon their decay to the ground state they emit (blue) light, which is amplified with a photo-multiplier tube and recorded with a CCD. The number of emitted photons is directly proportional to the number of trapped electronhole pairs, which itself is directly proportional to the energy that was deposited in the phosphorus layer by the ionizing radiation. Because of this direct relation, an absolute calibration of the digitized detector count versus number of incident ionizing particles can be found, given that their type and energy is known. IPs are generally sensitive to all kinds of ionizing radiation; their application is therefore limited to scenarios, where the energy and the type of incident radiation is known, e.g., in spectrometers. Such applications often benefit from the high dynamic range ($> 10^5$, extendable via multiple scanning), the good spatial resolution (here around 75 μm) and the ease of use (scanning takes just few minutes).

After recording the signal and digitalizing the data with the scanner, the stored information in the IP can be fully erased by the same process as used during scanning. This enables the reuse of the IP, presenting another major advantage over CR39 detectors. For this work, the IP signal for protons was cross-calibrated with CR39 detectors, yielding calibration curves very similar to those known from literature [4, 6–8]. For X-rays we used Ref. [2].

B.2 Devices

B.2.1 Wide Angle Ion Spectrometer

The wide angle ion spectrometer (WASP, iWASP) is a central diagnostics tool employed in this thesis to characterize laser-driven sources. The device consists of a 40 mm thick steel slit with a 100 μm to 1 mm wide opening over a length of 100–200 mm. This slit cuts a fan-beam out of the laser-driven particle beam. Behind the slit, strong (up to 1 T) permanent magnets deflect charged particles depending on their type (charge-to-mass ratio and mass) and their energy. Behind the magnetic field, the free drift serves to further disperse different energies spatially, before recording them on the detector. Meanwhile, X-rays travel in a straight line to the detector.

The resolution of the spectrometer is determined by the width of the slit-projection, and the magnetic dispersion of the signal [9]. The approximate energy resolution $\Delta E_{kin}/E_{kin}$ of the WASP can be calculated by solving the Lorentz equation in the non-relativistic case and taking into account the source magnification in the detector plane [9]:

$$\frac{\Delta E_{kin}}{E_{kin}} = \frac{2s}{y(1 - (s/2y)^2)^2}. \tag{B.1}$$

Hereby $s = d(1 + b/g)$ is the projected slit width in the detector plane, d is the slit width, $b = L_B + D$ and $g = L_{t-s} + L_s$. The second parameter is the deflection by the magnetic field

$$y = \frac{qBL_B(D + 0.5L_B)}{(2m_i E_{kin})^{0.5}}, \tag{B.2}$$

where q and m_i are the ion charge and mass respectively. In the current work we used spectrometer designs to support reasonable resolutions around 5–10% at 10 MeV proton kinetic energy, while maintaining good signal-to-noise ratio (i.e. slit widths).

In order to record unambiguous proton spectra, we used IP detectors covered all-over with 30–100 μm of Aluminum, producing a lower cutoff line because a certain minimum amount of energy is required for protons to pass this protection layer. Of course, at sufficiently high energy (higher than for protons), heavier ion contaminants in the ion beam could as well traverse the Aluminum protection layer and contribute to the recorded spectrum. In order to avoid this effect, and hence to record only protons, we added an additional layer of CR39 on top of some parts of the IP detector. In the observed energy ranges, this allowed to avoid heavier ion

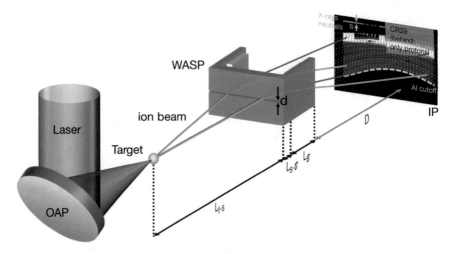

Fig. B.1 Wide angle ion spectrometer. The WASP consists of an entry slit at a distance L_{t-s} from the target and with thickness L_s, a dipole magnet of length L_B, a drift length D and a plane detector. Here it is used to record angularly resolved quantitative proton kinetic energy distributions

contaminations in the recorded spectrum, and thus to record unambiguous proton spectra.

In order to quantitatively analyze the recorded data, we performed a numerical particle tracing through the experimental setup for varying ion kinetic energies and angles (dots on the detector in Fig. B.1) including the relativistic effects. Besides the geometrical data, the most important input is the measured magnetic field, which is measured with a Hall sensor (Metrolab, THM 1176-HFC-PC, [10]) in a set of three-dimensional magnetic field components, within a three-dimensional geometric raster.

Finally, the calibration of the IP signal (PSL) to absolute proton numbers at a given energy was accomplished by cross-calibration with CR39 detectors at a set of energies and particle numbers and relying on the established linear response of the IP to deposited dose in the sensitive layer (cf. [5]). The calibration was then extrapolated to further energies using SRIM/TRIM calculations to evaluate the deposited energy per proton in the sensitive layer for a given kinetic energy. Overall, the calibration curve PSL/proton(E_{kin}) obtained is similar to calibrations in literature [6–8].

B.2.2   X-ray Cleaner

For the recording of bi-modal images we modified the WASP to work as an X-ray cleaner. In this context, the purpose of the detector is dedicated for the X-rays instead of for the ions; the magnetic field is used to deflect charged particle beams away from the directly projected X-ray image. In order to provide reasonable field of view, the slit-width was increased to 20 mm. The maximum slit-width (and with it the field of view for the X-ray image) is limited by the deflection y of the maximum energy (both for electrons and for ions), $y(E_{kin}^{max}) > s$. Only then, the recorded image will consist only of X-ray signal alone and not be contaminated by ion or electron signal. In the current setup with a magnetic field of only 200 mm length and 500 mm drift, proton energies of up to 25 MeV and electrons up to 200 MeV were deflected away from the X-ray image.

B.2.3 Thomson Parabola spectrometer

In the Thomson parabola spectrometer (TP, Fig. B.2), the slit of the wide-angle ion spectrometer is replaced by an entry pinhole of diameter d at a distance L_{t-p} from the target.

At distances of $L_{p-\varepsilon}$ and L_{p-B} the pinhole is followed by parallel (or anti-parallel) electric and magnetic fields of lengths L_ε and L_B. In contrast to the WASP, the Thomson parabola allows to record unambiguous ion spectra for multiple species (more precisely, ions of different charge-to-mass ratio) at once, at cost of the 1D spatial resolution. Again, in the non-relativistic case the ion motion can be calculated via the Lorentz equation. The magnetic deflection is till given by Eq. (B.2). The deflection by the electric field is

$$x = \frac{q\mathcal{E}L_\varepsilon(D_\varepsilon + 0.5L_\varepsilon)}{2E_{kin}}. \tag{B.3}$$

and naturally occurs perpendicular to the magnetic deflection. Both deflections together account for the typical parabolic shape of the recorded curves

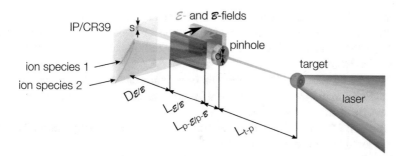

Fig. B.2 Thomson parabola spectrometer, TP. The TP consists of an entrance pinhole of width d at a distance L_{t-p} from the target, followed by parallel (or anti-parallel) electric and magnetic fields. The distance from pinhole to the electric and magnetic fields $L_{p-\varepsilon/p-B}$, the lengths of the fields $L_{\mathcal{E}/B}$ and the corresponding drift distances $D_{\mathcal{E}/B}$ can be adjusted for both fields separately

$$y^2 = x \frac{q\mathcal{B}^2 L_{\mathcal{B}}^2 (D_{\mathcal{B}} + 0.5L_{\mathcal{B}})^2}{m_i \mathcal{E} L_{\mathcal{E}} (D_{\mathcal{E}} + 0.5L_{\mathcal{E}})}. \tag{B.4}$$

The energy resolution of the TP is given by Eq. (B.1).

The separation of charge-to-mass-ratios is qualified by the energy, where two traces of different species will merge due to the finite size of the pinhole projection on the detector screen. This energy is [1]

$$E_M = \frac{q_1 \mathcal{E} L_{\mathcal{E}} (D_{\mathcal{E}} + 0.5L_{\mathcal{E}})}{s R_Q}, \tag{B.5}$$

with $R_Q = |(q_1/m_1 + q_2/m_2)|/|(q_1/m_1 - q_2/m_2)|$ and E_M is the merging energy for the ion species of charge q_1.

Per design, the TP employed in this work was able to separate the proton trace from the next higher trace (C^{6+}) up to 70 MeV, which was not reached in the experiment.

References

1. Jung D (2012) Ion acceleration from relativistic laser nano-target interaction. PhD thesis. Ludwig-Maximilians-Universität, München
2. Meadowcroft AL, Bentley CD, Stott EN (2008) Evaluation of the sensitivity and fading characteristics of an image plate system for x-ray diagnostics. Rev Sci Instrum 79(11):113102
3. Gautier DC et al (2008) A simple apparatus for quick qualitative analysis of CR39 nuclear track detectors. Rev Sci Instrum 79(10):10E536
4. Paudel Y et al (2011) CR39 imaging technique for quick track analysis of particles generated in high-intensity laser target interactions. J Instrum 6(08):T08004–T08004
5. Reinhardt S (2012) Detection of laser-accelerated protons. PhD thesis. LMU
6. Mančić A et al (2008) Absolute calibration of photostimulable image plate detectors used as (0.5–20 MeV) high-energy proton detectors. Rev Sci Instrum 79(7):073301
7. Freeman CG et al (2011) Calibration of a Thomson parabola ion spectrometer and Fujifilm imaging plate detectors for protons, deuterons, and alpha particles. Rev Sci Instrum 82(7):073301
8. Morrison JT (2013) Selective deuteron acceleration using target normal sheath acceleration. PhD thesis. Ohio State University
9. Jung D et al (2011) A novel high resolution ion wide angle spectrometer. Rev Sci Instrum 82(4):043301
10. MetroLab, https://www.metrolab.com/products/thm1176/

Appendix C
Theory of Radiographic Imaging with X-Rays and Ions

Parts of this thesis are concerned with the application of laser driven micro-sources to a variation of imaging applications. Here, brief introductions to radiographic imaging with a focus on basic theory for understanding of the experiments will be given. The first part of this section covers X-ray attenuation imaging and phase-contrast imaging. The second part of this section deals with the stopping of fast ions in matter as used in imaging applications.

C.1 Basics of X-ray Imaging

When X-rays penetrate through matter they are attenuated by several effects, namely photoelectric effect, Compton scattering and pair production. In the photoelectric effect [9–11] an X-ray photon of energy $E_{ph} = hc/\lambda$ is absorbed and ionizes an atom to generate a free electron with energy $E_{el} = E_{ph} - \Phi$, where Φ in this context denotes the ionization potential for the specific electronic state of the atom. This mechanism leads to material-characteristic discontinuous attenuation-edges in the spectrum. At these edges attenuation increases suddenly towards higher photon energies, when the higher photon energy suffices to ionize a quantum state with a larger work-function Φ of the atom. Compton scattering [12, 13] is the inelastic scattering of a photon by a quasi-free electron. As a result of Compton scattering, the photon energy will be reduced and parts of its energy will be transferred to the electron, where both the electron and the photon energy are determined by the initial photon and electron energies and the scattering angle. The third process, pair production [14, 15] in the electron or nuclear field, occurs only for photon energies beyond 1 MeV where energy is sufficient to create pairs of electrons and positrons, both of which have a rest-mass of $m_e c^2 = 0.511 \, \text{MeV}/c^2$. Depending on the photon energy, the absorption process can be dominated by either of those processes. Notably, the elastic Rayleigh scattering is always orders weaker than any of the other processes and thus negligible.

© Springer Nature Switzerland AG 2019
T. Ostermayr, *Relativistically Intense Laser–Microplasma Interactions*,
Springer Theses, https://doi.org/10.1007/978-3-030-22208-6

The overall effect of X-ray attenuation can be described by the Beer-Lambert [16–18] law

$$I_f = I_0 \exp\left(-\int \mu^\rho_{E_{ph}}(x)\mathrm{d}x\right),\tag{C.1}$$

where I_f is the final X-ray intensity based on incoming X-ray intensity I_0, x is the propagation direction of the X-ray beam and $\mu^\rho_{E_{ph}}(x)$ is the local attenuation coefficient, which itself depends on material, photon-energy E_{ph} and material-density ρ. The total absorption coefficient $\mu^\rho_{E_{ph}}$ is the sum of the individual absorption coefficients for the above mentioned processes. Already in the first series of papers by Röntgen [1–3] describing his discovery of X-rays, this material-, density- and thickness-dependence of X-ray attenuation was used for imaging of human tissue, e.g. to distinguish soft tissue from bones. The integrated absorption coefficient along a path through the object can be rewritten [5] as

$$p = \ln\frac{I_f}{I_0} = -\int \mu^\rho_{E_{ph}}(x)\mathrm{d}x.\tag{C.2}$$

Here Eq. (C.2) represents the Radon transform of the absorption coefficient of the material and can thus be retrieved in a computerized tomography. Analytic expressions can approximate the absorption coefficient for each of the three major processes depending on E_{ph}, Z and ρ, thereby connecting the measurement with the physical properties of the sample. E.g. the photoelectric effect, which dominates at low photon energies such as used here, roughly [5] goes as $\mu^\rho_{E_{ph}}(E) = \text{const.} \cdot Z^3/E^3_{ph}$. Of course, a closer look reveals that discontinuities at absorption edges are not modeled here. Empirical data or refined models are more accurate. However, it gives a good idea about how contrast in radiographic images is produced, and how it changes with photon energy or imaged material. An issue directly recognized in this scaling is the beam hardening effect that occurs when using a non-monochromatic source: since the absorption coefficient decreases with photon energy, lower energy photons will lose more intensity when penetrating through matter. The output spectrum is 'hardened', i.e. its spectral distribution is shifted towards higher photon energies. This needs to be taken into account for the interpretation of X-ray images and reconstruction of CT.

C.1.1 Projection Imaging

The use of X-rays in a point-projection scheme for microscopy has been used since more than half a century. In this scenario, the object is typically placed close, say at a distance L to a point-like source and thus geometrically magnified onto a detector at a distance D from the object by a factor of $M = (L + D)/L$. The resolution is then limited by the detector resolution or non-ideal source distribution. Most of the early advances in projection microscopy were driven by the parallel developments of electron diffraction microscopy, where finely tuned electron beams were developed that could be shined on anodes to produce small sources. Using such approaches allowed spatial resolutions down to the 0.1 μm level in the 1990s, and a number of problems was identified when going to smaller spatial scales. First, the spreading of electrons in the anode material is a problem if too large electron energies are used, since this intrinsically leads to increased source-distributions. Secondly, the final electron lens needed to be designed in a way to avoid aberrations, leading to low intensity and thus long exposure times. And thirdly, Fresnel diffraction led to image blurring in very-high resolution images. Besides other developments this latter one has raised great attention since it can indeed be turned into an advantage when introduced deliberately.

C.1.2 Phase Contrast

If we take one step back and consider the electro-magnetic-wave nature of X-rays by evaluating their electric fields rather than just counting photons (or intensities), we can use this property to create the radiographic contrast as well.[1] The scalar wave-function in a medium can be written as

$$\psi = \mathcal{E}_0 e^{inkx}, \tag{C.3}$$

where $\mathcal{E}_0$ is the amplitude of the field, $n = 1 - \delta + i\beta$ is the complex refractive index and k is the angular wavenumber. We can identify $2k\beta = \mu^\rho_{E_{ph}}$ with the attenuation from earlier. We can thus write

$$\psi = \mathcal{E}_0 e^{i(1-\delta)kx} e^{-\mu^\rho_{E_{ph}} x/2}. \tag{C.4}$$

While the first exponential function contains information on the phase, the second exponential models the attenuation. The total phase of the wave after passing through a medium of thickness D with respect to an undisturbed wave running through vacuum is

$$\phi = \frac{2\pi}{\lambda} \int_0^D \delta(x) dx, \tag{C.5}$$

where $\delta(x) = 2\pi\rho Z r_e k^{-2}$ is the local decrement of the real part of the refractive index, and $r_e = (4\pi\varepsilon_0)^{-1} e^2 m_e^{-1} c^{-2}$ is the classical electron radius. The phase-shift has several properties making it appealing for imaging applications; its magnitude is fundamentally independent from absorption in the material and thus dose on patients can be reduced. Similarly, objects that barely attenuate the X-ray beam can be made visible. Particularly the scaling with the wave-vector k is better than that for attenuation towards higher photon energies. Overall, the effect of phase-shift in the parameter-range relevant to medical imaging is typically around three orders of

---

[1]This section mostly follows arguments given in Ref. [8].

magnitude larger than that of attenuation. With the phase-shift of the light penetrating through a sample manifesting in the electric fields rather than in the direct intensity distribution, the phase contrast is usually measured with quasi-interferometric techniques. Directly behind the imaged object, in a plane that we denote $x = x_0 = 0$, the wave-function will read

$$\psi(0) = \mathcal{E}_0 e^{ikx_0} e^{i\phi + p/2}. \tag{C.6}$$

It directly follows that $I(0) = |\psi|^2 = \mathcal{E}_0^2 e^p$, with p defined in Eq. (C.2), and hence no phase-contrast will be measured in the intensity distribution close to the object. However, phase-contrast will be produced upon further propagation via Fresnel diffraction.

The Fresnel diffraction can well be calculated in the Fourier domain [8]. Here we focus on results of such analysis that are of relevance for experimental design and analysis. For a parallel incoming X-ray beam the analysis yields, that the feature-size in the object producing optimum phase contrast for a given object-detector distance D is [7, 8]

$$a = \sqrt{2\lambda D}. \tag{C.7}$$

In this configuration, the pure phase image will show, without absorption-contrast being present. The same discussion as presented above for spherical waves, as used in the point-projection imaging, yields a quite similar condition for the optimum conditions to observe phase-contrast. Consider an image of magnification $M = (D + L)/L$ with L and D being the source-object and object-detector distances respectively. Then the optimum feature size a_M is [8]

$$a_M = \sqrt{\frac{2\lambda D}{M}}. \tag{C.8}$$

Practically, a detector with pixel-size Δs will resolve phase-contrast if $\Delta s < a \cdot M$. The actual value measured by the detector is an intensity distribution; measuring the intensity distribution at a plane $x = D$ and another plane $x = 0$ allows to connect the measurement to physical values via [8]

$$I(D) - I(0) = -\frac{D\lambda}{2\pi} \nabla_T (I(0)\nabla_T \phi). \tag{C.9}$$

The connection of ϕ to physical properties of the sample is given in Eq. (C.5). Iterative algorithms allow to retrieve similar quantities from recordings in a single image plane. An interesting insight is, that the wavelength λ is, as a good approximation, separable in Eq. (C.9), meaning that regardless of the X-ray wavelength the intensity distribution will be similar except for a factor, in principle allowing for polychromatic phase-contrast imaging [4].

Descriptions above are given for ideal conditions for absorption imaging ($x = 0$) and ideal phase-contrast imaging, where the object in both cases is typically recognizable from the recorded intensity distribution. Yet it is noteworthy that more complex phase-sensitive imaging modes exist when going beyond the condition for ideal phase-contrast imaging, i.e. increasing object-detector distance D for a given feature size d. These are referred to as holographic and as far-field (or Fraunhofer) regime. The latter is well known in context with X-ray crystallography. In imaging experiments, objects typically contain a range of feature size and therefore it is often the case that in a specific configuration some features show in direct phase contrast imaging while smaller features show in a holographic mode.

There is one more crucial consideration for phase-contrast imaging in the point-projection scheme when it comes to experiments. That is the assumption of a perfectly point-like source above. Or in other words, the effects of a finite spatial source distribution and consequent partial coherence. Such will lead to reduced or even invisible contrast towards higher spatial frequencies [8]. As a rule of thumb, phase contrast will be visible as long as the transverse coherence length in the object plane $d_c = \lambda L / \sigma$ is equal or larger than the first Fresnel zone at the same location, $r_F = (\lambda D / M)^{1/2}$. In other words, the projected source size σ must be smaller than the projection of the optimum feature size, Ma_M, in order not to blur the image.

C.2   Basics of Ion Stopping and Ion Radiographic Imaging

Ion radiographic imaging relies on the stopping characteristics of ion beams penetrating through matter.[2] The electronic energy loss dominates their stopping; in single collisions inside the sample, the energetic ion can cause atomic excitation, collective excitation, or ionization and thereby lose a part of its energy, typically less than 100 eV per collision. Due to the statistical nature of these processes, a thin medium will generally lead to a small number of collisions and a large variance in the total energy loss behind the medium.

Ion stopping in a wide range of energies ($0.1 < \beta\gamma < 1000$) which are relevant to radiotherapy and radiographic imaging can be described by the Bethe-Bloch formula [20, 21]:

$$-\left\langle \frac{dE}{dx} \right\rangle = \rho \frac{4\pi e^4}{m_e v^2 u} \frac{Z}{A} z^2 \left[\ln \frac{2m_e v^2}{\iota} + \ln \frac{1}{1-\beta^2} - \beta^2 - \frac{C}{Z} - \frac{\delta}{2} \right], \quad \text{(C.10)}$$

where

$$\frac{4\pi e^4}{m_e v^2 u} \frac{Z}{A} z^2 = 0.307075 \frac{z^2}{\beta^2} \frac{Z}{A} \quad \text{(C.11)}$$

is given in units of MeV cm/g, e is the electron charge, u is the atomic mass unit, z is the charge of the ion, m_e is the mass of the electron, v is the velocity of the ion and $\beta = v/c$. Z and A are the charge and atomic mass of the target material, ι is the mean excitation energy of the target material, C/Z and $\delta/2$ are shell corrections

---

[2]The description of ion stopping relies on descriptions given in Ref. [19].

and density-effect corrections respectively. The Bethe formula describes average ion stopping in terms of particle and material properties. An example of the latter is depicted in Fig. 1.3, showing that protons lose much of their energy in close vicinity to the point where they are stopped, in contrast to X-rays which lose intensity exponentially when traveling through matter. The location of maximum energy loss (maximum dose deposition) is known as the Bragg peak. Medical applications utilize the localized dose deposition of ions for their use in charged particle therapy of cancers, where the spatial dose-confinement enables to reduce dose on healthy tissues surrounding the tumor.

In order to estimate the range of a proton beam in matter, the so-called continuous slowing down approximation (CSDA) is often used [22]. This simply integrates the (inverse) average stopping to calculate the mean range for particles of a given energy in the material

$$R_{CSDA} = \int_0^{E_0} -\left\langle \frac{dE}{dx} \right\rangle^{-1} dE. \tag{C.12}$$

C.2.1 Energy and Range Straggling

The above description is given for an ensemble average. This directly implies the statistical nature of the ion-stopping when considering the high particle fluence used in charged particle therapy or ion radiography. Effects of these statistical processes on the ion beam penetrating through a sample will be discussed shortly. Considering an initially perfectly mono-energetic (pencil)beam entering the sample material, statistical fluctuations of the energy-loss process (which relies on single collisions) broaden its spectral bandwidth. Of course, this energy straggling goes along with a range-straggling. This effect is a fundamental limit for the density resolution of ion radiographies and tomographies.

A variety of models exists to describe the effect in dependence of the target thickness. For most applications in ion beam therapy and imaging, Bohrs theory for thick layers of matter provides a good enough estimate: in the limit of many collisions, the energy and range distributions become Gaussian [23, 24]

$$f(\Delta_{E,R}) = \frac{1}{\sqrt{2\pi\sigma_{E,R}^2}} \exp\left[-\frac{(\Delta_{E,R} - \overline{\Delta_{E,R}})^2}{2\sigma_{E,R}^2}\right]. \tag{C.13}$$

The variance of the range straggling σ_R^2 is connected to the variance of the energy loss σ_E^2 via [25]

$$\sigma_R^2 = \int_0^{E_0} \frac{d\sigma_E}{dx} \left\langle \frac{dE}{dx} \right\rangle^{-3} dE. \tag{C.14}$$

It is rather common to use a power-law approximation for the width of the range straggling in terms of the ion beam range and the ion mass number [6, 26, 27]

$$\sigma_R[\text{cm}] = k_w R_w^m = 0.012 \cdot A^{-0.5} R_w^{0.935}, \tag{C.15}$$

where k_w is a material constant, m is empirically determined, A is the ion's mass number and R_w is the range of the ion beam (given its initial energy E_0) in water. Strictly this approximation is good only for $2 < R_w < 40$ cm. The dependence on A and R_w shows that—for a given range—heavier ion species are less susceptible to energy and range straggling, and therefore generally beneficial for depth (or density) resolution in imaging applications. Similarly, the range straggling grows about linear with the range itself, and the relative thickness (or density) resolution in radiography is therefore almost constant.

C.2.2 Multiple Coulomb Scattering

Besides the energy and range straggling, the stochastic nature of charged particle interactions also causes small lateral deflections from the ion's original path. These are predominantly caused by elastic Coulomb interactions via the repulsive forces from the sample's nuclei, the so-called Multi Coulomb Scattering (MCS). While these Coulomb interactions with atomic nuclei do barely contribute to the overall energy-loss, they are still of great importance for dosimetry and transmission imaging, as they cause the lateral broadening the (pencil)beam and thereby limit the lateral resolution. An analytical description for the distribution of the scattering angles θ of ions after the passage through a sample of thickness L was found in Ref. [28]. For small deflection angles ($<10°$), higher-order terms in Molire's solution can be neglected and—again—yield a simple Gaussian distribution [28, 29]

$$f(\theta) = \frac{1}{\sqrt{2\pi\sigma_\theta^2}} \exp\left[-\frac{\theta^2}{2\sigma_\theta^2}\right]. \tag{C.16}$$

The width of the Gaussian function σ_θ, is given by Highland's formula as [25, 30]

$$\sigma_\theta[\text{rad}] = \frac{14.1\text{MeV}}{\beta pc} z \sqrt{\frac{L}{L_{rad}}} \left[1 + \frac{1}{9} \cdot \log_{10}\left(\frac{L}{L_{rad}}\right)\right]. \tag{C.17}$$

And once again, it is useful to estimate the effects of MCS with a power-law [26]

$$\sigma_y = 0.0294 \cdot R_w^{0.896} z^{-0.207} A^{-0.396}, \tag{C.18}$$

where σ_y is the lateral width of an initial pencil beam after penetration of the material in cm. It becomes obvious that straggling gains significance with the range traveled in the material, while increased charge z and mass A of the incident ion could reduce the effect. Again, the effect grows almost linearly with traveled range.

C.2.3 Nuclear Interactions

Up to here, electromagnetic processes were considered as the main cause for energy loss and straggling of the incident primary particles. In addition, ion beams can also undergo nuclear reactions in the target material. The proton reaction happens at the nucleon level and is modeled with the exciton formalism [31]; protons, neutrons and light fragments are emitted, and the residual nucleus approaches an equilibrium state with certain remaining excitation energy [32].

The main consequence of nuclear interactions is the exponential reduction of the beam fluence in the sample, which is relevant for particle counting based spectral evaluations as considered herein and requires proper modeling. As example, considering a proton ion beam with 16 cm range in water, only about 80% of the originally impinging protons reach the Bragg Peak. Generally, due to this effect an increase in penetration depth is associated with a reduced ratio of the Bragg-Peak to the Plateau region before it. Thus, the unambiguous detection of residual particle energy via residual range in the detector requires good Signal-to-Noise ratio. For protons penetrating into water, the most significant contribution of nuclear interactions is the production of neutrons, knocked out of the target material. These do interact weakly and can deliver dose well beyond the Bragg-peak of the original proton beam, which is significant in considerations for charged particle therapy. For a particle-counting based spectrum evaluation as used in this work, it is necessary to include these effects in the simulations that are used to calibrate the particle-count versus the object thickness.

References

1. Röntgen WC (1895) Ueber eine neue Art von Strahlen. Vorläufige Mitteilung. In: Sitzungsberichte der Physikalisch-Medicinischen Gesellschaft zu Würzburg 9. Würzburg: Verlag und Druck der Stahel'schen K. Hof- und Universitäts-Buch- und Kunsthandlung, pp 132–141
2. Röntgen WC (1896) Ueber eine neue Art von Strahlen. 2. Mitteilung. In: Sitzungsberichte der Physikalisch-Medicinischen Gesellschaft zu WPürzburg. Würzburg: Verlag und Druck der Stahel'schen K. Hof- und Universitäts-Buch- und Kunsthandlung, pp 1–9
3. Röntgen WC (1897) Weitere Beobachtungen über die Eigenschaften der X-Strahlen. In: vol. Band Erster Halbband. Sitzungsberichte der Königlich Preußischen Akademie der Wissenschaften zu Berlin. Verl. d. Kgl. Akad. d. Wiss., Berlin, pp 576–592
4. Wilkins SW et al (1996) Phase-contrast imaging using polychromatic hard Xrays. Nature 384(6607):335–338
5. Jerrold T Bushberg et al (2012) The essential physics of medical imaging, 3rd edn. Lippincott Williams and Wilkins, Philadelphia, PA

6. Newhauser WD, Zhang R (2015) The physics of proton therapy. Phys Med Biol 60(8):R155

7. Cloetens P et al (1997) Observation of microstructure and damage in materials by phase sensitive radiography and tomography. J Appl Phys 81(9):5878

8. Mayo SC et al (2002) Quantitative X-ray projection microscopy: phase-contrast and multi-spectral imaging. J Microsc 207.Pt 2, pp 79–96

9. Einstein A (1905) Über einen die Erzeugung und Verwandlung des Lichtes betreffenden heuristischen Gesichtspunkt. Annalen Phys 17

10. Hertz H (1887) Ueber einen Ein uss des ultravioletten Lichtes auf die electrische Entladung. Annalen Phys 267(8):983–1000

11. Lenard P (1902) Ueber die lichtelektrische Wirkung. Annalen Phys 313(5):149–198

12. Compton AH (1923) A quantum theory of the scattering of X-rays by light elements. Phys Rev 21:483–502

13. Compton AH (1925) Simon AW Directed quanta of scattered X-Rays. Phys Rev 26:289–299

14. Anderson CD (1933) The positive electron. Phys Rev 43:491–494

15. Motz JW, Olsen HA, Koch HW (1969) Pair production by photons. Rev Mod Phys 41:581–639

16. Bouguer P (1729) Essai d'optique, Sur la gradation de la lumière. Claude Jombert, 164 ff

17. Lambert JH (1760) Photometria, sive de mensura et gradibus luminis, colorum et umbrae. Sumptipus Vidae Eberhardi Klett, Augsburg

18. Beer A (1852) Bestimmung der Absorption des rothen Lichts in farbigen Flüssigkeiten. Annalen Phys Chem 86:78–88

19. Particle Data Group (2016) Review of particle physics. Chin Phys C 40(10)

20. Bethe H (1930) Zur Theorie des Durchgangs schneller Korpuskularstrahlen durch Materie. Annalen Phys 397(3):325–400

21. Bloch F (1933) Zur Bremsung rasch bewegter Teilchen beim Durchgang von Materie. Annalen Phys 408(3):285–320

22. Carron NJ (2006) An introduction to the passage of energetic particles through matter. Taylor & Francis

23. Bohr N (1940) Scattering and stopping of fission fragments. Phys Rev 58:654–655

24. Lewis HW (1952) Range straggling of a nonrelativistic charged particle. Phys Rev 85:20–24

25. Schardt D, Elsässer T, Schulz-Ertner D (2010) Heavy-ion tumor therapy: physical and radiobiological benefits. Rev Mod Phys 82:383–425

26. Chu WT, Ludewigt BA, Renner TR (1993) Instrumentation for treatment of cancer using proton and lightion beams. Rev Sci Instrum 64(8):2055–2122

27. Bortfeld T (1997) An analytical approximation of the Bragg curve for therapeutic proton beams. Med Phys 24(12):2024–2033

28. Molière G (1947) Theorie der Streuung schneller geladener Teilchen I. Einzelstreuung am abgeschirmten Coulomb-Feld. Z Naturforschung Teil A 2:133–145

29. Molière G (1948) Theorie der Streuung schneller geladener Teilchen II. Mehrfachund Vielfachstreuung. Z Naturforschung Teil A 3:78–97
30. Highland VL (1975) Some practical remarks on multiple scattering. Nucl Inst Methods 129(2):497–499
31. Griffin JJ (1966) Statistical model of intermediate structure. Phys Rev Lett 17:478–481
32. Kraan AC (2015) Range verification methods in particle therapy: underlying physics and monte carlo modeling. Front Oncol 5:150

Printed in the United States
By Bookmasters